P9-DFS-847

Learn, Practice, Succeed

Eureka Math®
Grade 8
Module 7

Published by Great Minds®.

Copyright © 2019 Great Minds®.

Printed in the U.S.A.

This book may be purchased from the publisher at eureka-math.org.

10 9 8 7 6 5 4 3 2

ISBN 978-1-64054-986-9

G8-M7-LPS-05.2019

Students, families, and educators:

Thank you for being part of the *Eureka Math®* community, where we celebrate the joy, wonder, and thrill of mathematics.

In *Eureka Math* classrooms, learning is activated through rich experiences and dialogue. That new knowledge is best retained when it is reinforced with intentional practice. The *Learn, Practice, Succeed* book puts in students' hands the problem sets and fluency exercises they need to express and consolidate their classroom learning and master grade-level mathematics. Once students learn and practice, they know they can succeed.

What is in the Learn, Practice, Succeed *book?*

Fluency Practice: Our printed fluency activities utilize the format we call a Sprint. Instead of rote recall, Sprints use patterns across a sequence of problems to engage students in reasoning and to reinforce number sense while building speed and accuracy. Sprints are inherently differentiated, with problems building from simple to complex. The tempo of the Sprint provides a low-stakes adrenaline boost that increases memory and automaticity.

Classwork: A carefully sequenced set of examples, exercises, and reflection questions support students' in-class experiences and dialogue. Having classwork preprinted makes efficient use of class time and provides a written record that students can refer to later.

Exit Tickets: Students show teachers what they know through their work on the daily Exit Ticket. This check for understanding provides teachers with valuable real-time evidence of the efficacy of that day's instruction, giving critical insight into where to focus next.

Homework Helpers and Problem Sets: The daily Problem Set gives students additional and varied practice and can be used as differentiated practice or homework. A set of worked examples, Homework Helpers, support students' work on the Problem Set by illustrating the modeling and reasoning the curriculum uses to build understanding of the concepts the lesson addresses.

Homework Helpers and Problem Sets from prior grades or modules can be leveraged to build foundational skills. When coupled with *Affirm®, Eureka Math*'s digital assessment system, these Problem Sets enable educators to give targeted practice and to assess student progress. Alignment with the mathematical models and language used across *Eureka Math* ensures that students notice the connections and relevance to their daily instruction, whether they are working on foundational skills or getting extra practice on the current topic.

Where can I learn more about Eureka Math *resources?*

The Great Minds® team is committed to supporting students, families, and educators with an ever-growing library of resources, available at eureka-math.org. The website also offers inspiring stories of success in the *Eureka Math* community. Share your insights and accomplishments with fellow users by becoming a *Eureka Math* Champion.

Best wishes for a year filled with "aha" moments!

Jill Diniz

Jill Diniz
Chief Academic Officer, Mathematics
Great Minds

Contents

Module 7: Introduction to Irrational Numbers Using Geometry

Note: The figures in this lesson are not drawn to scale.

Example 1

Write an equation that allows you to determine the length of the unknown side of the right triangle.

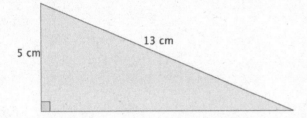

Example 2

Write an equation that allows you to determine the length of the unknown side of the right triangle.

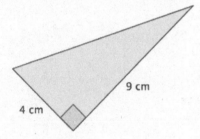

Example 3

Write an equation to determine the length of the unknown side of the right triangle.

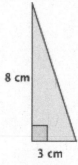

8 cm

3 cm

Example 4

In the figure below, we have an equilateral triangle with a height of 10 inches. What do we know about an equilateral triangle?

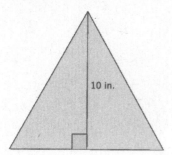

10 in.

Exercises

1. Use the Pythagorean theorem to find a whole number estimate of the length of the unknown side of the right triangle. Explain why your estimate makes sense.

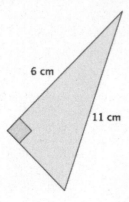

6 cm

11 cm

2. Use the Pythagorean theorem to find a whole number estimate of the length of the unknown side of the right triangle. Explain why your estimate makes sense.

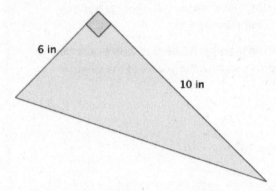

3. Use the Pythagorean theorem to find a whole number estimate of the length of the unknown side of the right triangle. Explain why your estimate makes sense.

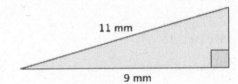

Lesson Summary

Perfect square numbers are those that are a product of an integer factor multiplied by itself. For example, the number 25 is a perfect square number because it is the product of 5 multiplied by 5.

When the square of the length of an unknown side of a right triangle is not equal to a perfect square, you can estimate the length as a whole number by determining which two perfect squares the square of the length is between.

Example:

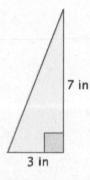

7 in

3 in

Let c in. represent the length of the hypotenuse. Then,

$$3^2 + 7^2 = c^2$$
$$9 + 49 = c^2$$
$$58 = c^2.$$

The number 58 is not a perfect square, but it is between the perfect squares 49 and 64. Therefore, the length of the hypotenuse is between 7 in. and 8 in. but closer to 8 in. because 58 is closer to the perfect square 64 than it is to the perfect square 49.

EUREKA
MATH

Name _____   Date _____

1. Determine the length of the unknown side of the right triangle. If you cannot determine the length exactly, then
    determine which two integers the length is between and the integer to which it is closest.

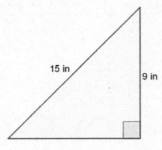

2. Determine the length of the unknown side of the right triangle. If you cannot determine the length exactly, then
    determine which two integers the length is between and the integer to which it is closest.

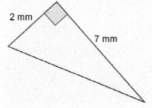

1. Use the Pythagorean theorem to estimate the length of the unknown side of the right triangle. Explain why your estimate makes sense.

> I remember the Pythagorean theorem from previous lessons.

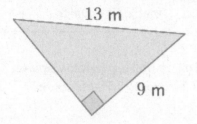

13 m

9 m

Let x m represent the length of the unknown side.

$$9^2 + x^2 = 13^2$$
$$81 + x^2 = 169$$
$$x^2 = 88$$

> The number 88 is not a perfect square, but I do know that 81 and 100 are perfect squares, and 88 is between them.

The number 88 is between the perfect squares 81 and 100. Since 88 is closer to 81 than it is to 100, the length of the unknown side of the triangle is closer to 9 m than it is to 10 m.

2. Use the Pythagorean theorem to estimate the length of the unknown side of the right triangle. Explain why your estimate makes sense.

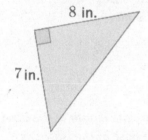

8 in.

7 in.

Let c in. represent the length of the hypotenuse.

$$7^2 + 8^2 = c^2$$
$$49 + 64 = c^2$$
$$113 = c^2$$

The number 113 is between the perfect squares 100 and 121. Since 113 is closer to 121 than it is to 100, the length of the hypotenuse of the triangle is closer to 11 in. than it is to 10 in.

3. Use the Pythagorean theorem to estimate the length of the unknown side of the right triangle. Explain why your estimate makes sense.

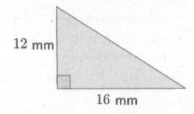

Let c mm represent the length of the hypotenuse.

$$12^2 + 16^2 = c^2$$
$$144 + 256 = c^2$$
$$400 = c^2$$
$$20 = c$$

The number 400 is a perfect square, so I know that 20 is equal to c.

The length of the hypotenuse is 20 mm. The Pythagorean theorem led me to the fact that the square of the unknown side is 400. We know 400 is a perfect square, and 400 is equal to 20^2; therefore, $c = 20$, and the length of the hypotenuse of the triangle is 20 mm.

4. The triangle below is an isosceles triangle. Use what you know about the Pythagorean theorem to determine the approximate length of the base of the isosceles triangle.

I can find the length of the base of one triangle and multiply that number by 2 since the two right triangles are congruent.

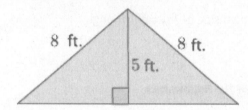

Let x ft. represent the length of the base of one of the right triangles of the isosceles triangle.

$$x^2 + 5^2 = 8^2$$
$$x^2 + 25 = 64$$
$$x^2 = 39$$

Since 39 is between the perfect squares 36 and 49 but closer to 36, the approximate length of the base of the right triangle is 6 ft. Since there are two right triangles, the length of the base of the isosceles triangle is approximately 12 ft.

© 2019 Great Minds®. eureka-math.org

EUREKA
MATH

5. Give an estimate for the area of the triangle shown below. Explain why it is a good estimate.

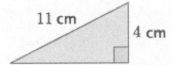

Let x cm represent the length of the base of the right triangle.

$$x^2 + 4^2 = 11^2$$

$$x^2 + 16 = 121$$

$$x^2 = 105$$

> I need to use the base, b, and height, h, of the triangle to find the area, A, of a triangle; $A = \frac{1}{2}bh$.

Since 105 is between the perfect squares 100 and 121 but closer to 100, the approximate length of the base is 10 cm. $A = \frac{1}{2}(10)(4) = 20$. So, the approximate area of the triangle is 20 cm^2.

> The hypotenuse is the longest side, so the base must be less than 11 cm.

20 cm^2 is a good estimate because of the approximation of the length of the base. Furthermore, since the hypotenuse is the longest side of the right triangle, approximating the length of the base as 10 cm makes mathematical sense because it has to be shorter than the hypotenuse.

1. Use the Pythagorean theorem to estimate the length of the unknown side of the right triangle. Explain why your estimate makes sense.

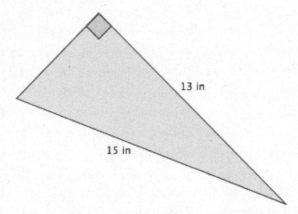

13 in

15 in

2. Use the Pythagorean theorem to estimate the length of the unknown side of the right triangle. Explain why your estimate makes sense.

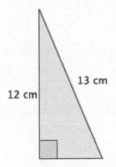

13 cm

12 cm

3. Use the Pythagorean theorem to estimate the length of the unknown side of the right triangle. Explain why your estimate makes sense.

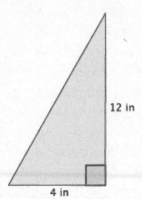

12 in

4 in

4. Use the Pythagorean theorem to estimate the length of the unknown side of the right triangle. Explain why your estimate makes sense.

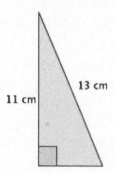

5. Use the Pythagorean theorem to estimate the length of the unknown side of the right triangle. Explain why your estimate makes sense.

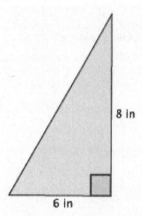

6. Determine the length of the unknown side of the right triangle. Explain how you know your answer is correct.

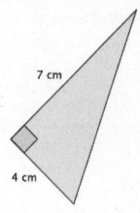

7. Use the Pythagorean theorem to estimate the length of the unknown side of the right triangle. Explain why your estimate makes sense.

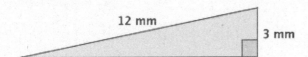

8. The triangle below is an isosceles triangle. Use what you know about the Pythagorean theorem to determine the approximate length of the base of the isosceles triangle.

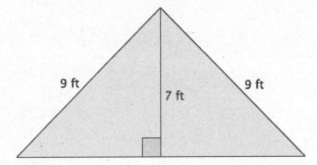

9. Give an estimate for the area of the triangle shown below. Explain why it is a good estimate.

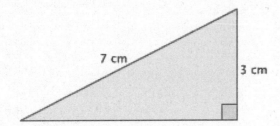

Exercises 1–4

1. Determine the positive square root of 81, if it exists. Explain.

2. Determine the positive square root of 225, if it exists. Explain.

3. Determine the positive square root of −36, if it exists. Explain.

4. Determine the positive square root of 49, if it exists. Explain.

Discussion

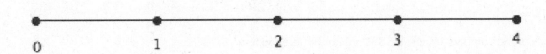

Exercises 5–9

Determine the positive square root of the number given. If the number is not a perfect square, determine which whole number the square root would be closest to, and then use *guess and check* to give an approximate answer to one or two decimal places.

5. $\sqrt{49}$

6. $\sqrt{62}$

7. $\sqrt{122}$

8. $\sqrt{400}$

9. Which of the numbers in Exercises 5–8 are not perfect squares? Explain.

EUREKA
MATH

Lesson Summary

A positive number whose square is equal to a positive number b is denoted by the symbol $\sqrt{b}$. The symbol $\sqrt{b}$ automatically denotes a positive number. For example, $\sqrt{4}$ is always 2, not -2. The number $\sqrt{b}$ is called a *positive square root of b*.

The square root of a perfect square of a whole number is that whole number. However, there are many whole numbers that are not perfect squares.

Name _____   Date _____

1. Write the positive square root of a number x in symbolic notation.

2. Determine the positive square root of 196. Explain.

3. The positive square root of 50 is not an integer. Which whole number does the value of $\sqrt{50}$ lie closest to? Explain.

4. Place the following numbers on the number line in approximately the correct positions: $\sqrt{16}$, $\sqrt{9}$, $\sqrt{11}$, and 3.5.

3   4

Determine the positive square root of the number given. If the number is not a perfect square, determine the integer to which the square root would be closest.

1. $\sqrt{196}$

    14

> I know that perfect squares have square roots that are equal to integers.

2. $\sqrt{225}$

    15

3. Between which two integers will $\sqrt{22}$ be located? Explain how you know.

    The number 22 is not a perfect square. It is between the perfect squares 16 and 25 but closer to 25. Therefore, the square root of 25 is between the integers 4 and 5 because $\sqrt{16} = 4$ and $\sqrt{25} = 5$ and $\sqrt{16} < \sqrt{22} < \sqrt{25}$.

4. Place the following list of numbers in their approximate locations on a number line.
    $$\sqrt{60}, \sqrt{38}, \sqrt{65}, \sqrt{90}, \sqrt{72}, \text{ and } \sqrt{51}$$

> I can use the same reasoning from Problem 2 to determine the placement of the square roots on the number line.

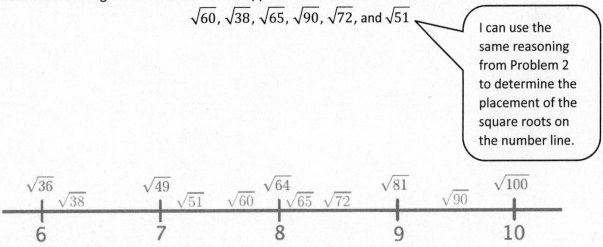

Answers are noted in blue.

Determine the positive square root of the number given. If the number is not a perfect square, determine the integer to which the square root would be closest.

1. $\sqrt{169}$

2. $\sqrt{256}$

3. $\sqrt{81}$

4. $\sqrt{147}$

5. $\sqrt{8}$

6. Which of the numbers in Problems 1–5 are not perfect squares? Explain.

7. Place the following list of numbers in their approximate locations on a number line.

$$\sqrt{32}, \sqrt{12}, \sqrt{27}, \sqrt{18}, \sqrt{23}, \text{ and } \sqrt{50}$$

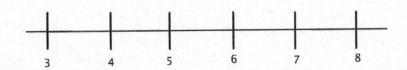

8. Between which two integers will $\sqrt{45}$ be located? Explain how you know.

Opening

The numbers in each column are related. Your goal is to determine how they are related, determine which numbers belong in the blank parts of the columns, and write an explanation for how you know the numbers belong there.

Find the Rule Part 1

1	1
2	
3	9
	81
11	121
15	
	49
10	
12	
	169
m	
	n

Find the Rule Part 2

1	1
2	
3	27
	125
6	216
11	
	64
10	
7	
	2,744
p	
	q

Exercises

Find the positive value of x that makes each equation true. Check your solution.

1. $x^2 = 169$

     a. Explain the first step in solving this equation.

     b. Solve the equation, and check your answer.

2. A square-shaped park has an area of 324 yd^2. What are the dimensions of the park? Write and solve an equation.

3. $625 = x^2$

4. A cube has a volume of 27 in^3. What is the measure of one of its sides? Write and solve an equation.

Lesson 3:   Existence and Uniqueness of Square Roots
                   and Cube Roots

EUREKA
MATH®

5. What positive value of x makes the following equation true: $x^2 = 64$? Explain.

6. What positive value of x makes the following equation true: $x^3 = 64$? Explain.

7. Find the positive value of x that makes the equation true: $x^2 = 256^{-1}$.

8. Find the positive value of x that makes the equation true: $x^3 = 343^{-1}$.

9. Is 6 a solution to the equation $x^2 - 4 = 5x$? Explain why or why not.

Lesson Summary

The symbol $\sqrt[n]{}$ is called a *radical*. An equation that contains that symbol is referred to as a *radical equation*. So far, we have only worked with square roots (i.e., $n = 2$). Technically, we would denote a positive square root as $\sqrt[2]{}$, but it is understood that the symbol $\sqrt{}$ alone represents a positive square root.

When $n = 3$, then the symbol $\sqrt[3]{}$ is used to denote the cube root of a number. Since $x^3 = x \cdot x \cdot x$, the cube root of x^3 is x (i.e., $\sqrt[3]{x^3} = x$).

The square or cube root of a positive number exists, and there can be only one positive square root or one cube root of the number.

Lesson 3:   Existence and Uniqueness of Square Roots
                    and Cube Roots

EUREKA MATH

Name _____   Date _____

Find the positive value of x that makes each equation true. Check your solution.

1. $x^2 = 225$

    a. Explain the first step in solving this equation.

    b. Solve and check your solution.

2. $x^3 = 64$

3. $x^2 = 361^{-1}$

4. $x^3 = 1000^{-1}$

1. A square-shaped park has an area of 900 ft^2. What are the dimensions of the park? Write and solve an equation.

   Let x ft. represent the length of one side of the park.

$$x^2 = 900$$
$$\sqrt{x^2} = \sqrt{900}$$
$$x = \sqrt{900}$$
$$x = 30$$

   Check:
$$30^2 = 900$$
$$900 = 900$$

   The square park is 30 ft. in length and 30 ft. in width.

2. A cube has a volume of 125 in^3. What is the measure of one of its sides? Write and solve an equation.

   Let x in. represent the length of one side of the cube.

   I need to use the cube root symbol, namely $\sqrt[3]{x}$, to find the cube root of a number x.

$$x^3 = 125$$
$$\sqrt[3]{x^3} = \sqrt[3]{125}$$
$$x = \sqrt[3]{125}$$
$$x = 5$$

   Check:
$$5^3 = 125$$
$$125 = 125$$

   The cube has a side length of 5 in.

3. Find the value of x that makes the equation true: $x^3 = 216^{-1}$.

$$x^3 = 216^{-1}$$

Check:

$$\sqrt[3]{x^3} = \sqrt[3]{216^{-1}}$$

$$(6^{-1})^3 = 216^{-1}$$

$$x = \sqrt[3]{216^{-1}}$$

$$6^{-3} = 216^{-1}$$

$$x = \sqrt[3]{\dfrac{1}{216}}$$

$$\dfrac{1}{6^3} = 216^{-1}$$

> I remember that I can rewrite 216^{-1} as $\dfrac{1}{216}$ to make the problem easier.

$$x = \dfrac{1}{6}$$

$$\dfrac{1}{216} = 216^{-1}$$

$$x = 6^{-1}$$

$$216^{-1} = 216^{-1}$$

4. Find the positive value of x that makes the following equation true: $x^2 - 22 = 99$.

$$x^2 - 22 = 99$$

$$x^2 - 22 + 22 = 99 + 22$$

$$x^2 = 121$$

$$\sqrt{x^2} = \sqrt{121}$$

$$x = 11$$

The positive value for x that makes the equation true is **11**.

Lesson 3:   Existence and Uniqueness of Square Roots
                and Cube Roots

EUREKA
MATH

Find the positive value of x that makes each equation true. Check your solution.

1. What positive value of x makes the following equation true: $x^2 = 289$? Explain.

2. A square-shaped park has an area of 400 yd^2. What are the dimensions of the park? Write and solve an equation.

3. A cube has a volume of 64 in^3. What is the measure of one of its sides? Write and solve an equation.

4. What positive value of x makes the following equation true: $125 = x^3$? Explain.

5. Find the positive value of x that makes the equation true: $x^2 = 441^{-1}$.
   a. Explain the first step in solving this equation.
   b. Solve and check your solution.

6. Find the positive value of x that makes the equation true: $x^3 = 125^{-1}$.

7. The area of a square is 196 in^2. What is the length of one side of the square? Write and solve an equation, and then check your solution.

8. The volume of a cube is 729 cm^3. What is the length of one side of the cube? Write and solve an equation, and then check your solution.

9. What positive value of x would make the following equation true: $19 + x^2 = 68$?

Opening Exercise

a.

   i.   What does $\sqrt{16}$ equal?

   ii.   What does 4×4 equal?

   iii. Does $\sqrt{16} = \sqrt{4 \times 4}$?

b.

   i.   What does $\sqrt{36}$ equal?

   ii.   What does 6×6 equal?

   iii. Does $\sqrt{36} = \sqrt{6 \times 6}$?

c.

   i.   What does $\sqrt{121}$ equal?

   ii.   What does 11×11 equal?

   iii. Does $\sqrt{121} = \sqrt{11 \times 11}$?

d.

   i.   What does $\sqrt{81}$ equal?

   ii.   What does 9×9 equal?

   iii. Does $\sqrt{81} = \sqrt{9 \times 9}$?

e. Rewrite $\sqrt{20}$ using at least one perfect square factor.

f. Rewrite $\sqrt{28}$ using at least one perfect square factor.

Example 1

Simplify the square root as much as possible.

$\sqrt{50} =$

Example 2

Simplify the square root as much as possible.

$\sqrt{28} =$

Exercises 1–4

Simplify the square roots as much as possible.

1. $\sqrt{18}$

2. $\sqrt{44}$

3. $\sqrt{169}$

4. $\sqrt{75}$

Lesson 4:   Simplifying Square Roots

EUREKA
MATH

Example 3

Simplify the square root as much as possible.

$\sqrt{128} =$

Example 4

Simplify the square root as much as possible.

$\sqrt{288} =$

Exercises 5–8

5. Simplify $\sqrt{108}$.

6. Simplify $\sqrt{250}$.

7. Simplify $\sqrt{200}$.

8. Simplify $\sqrt{504}$.

Lesson Summary

Square roots of some non-perfect squares can be simplified by using the factors of the number. Any perfect square factors of a number can be simplified.

For example:

$$\sqrt{72} = \sqrt{36 \times 2}$$
$$= \sqrt{36} \times \sqrt{2}$$
$$= \sqrt{6^2} \times \sqrt{2}$$
$$= 6 \times \sqrt{2}$$
$$= 6\sqrt{2}$$

Lesson 4: Simplifying Square Roots

EUREKA MATH®

Name _____   Date _____

Simplify the square roots as much as possible.

1. $\sqrt{24}$

2. $\sqrt{338}$

3. $\sqrt{196}$

4. $\sqrt{2420}$

Simplify each of the square roots in Problems 1–4 as much as possible.

1. $\sqrt{8}$

$$\sqrt{8} = \sqrt{2 \times 4}$$
$$= \sqrt{2 \times 2^2}$$
$$= 2 \times \sqrt{2}$$
$$= 2\sqrt{2}$$

> I need to rewrite 8 with factors of perfect squares.

2. $\sqrt{150}$

$$\sqrt{150} = \sqrt{2 \times 3 \times 5^2}$$
$$= \sqrt{2} \times \sqrt{3} \times \sqrt{5^2}$$
$$= 5 \times \sqrt{2 \times 3}$$
$$= 5\sqrt{6}$$

> I can rewrite $\sqrt{2} \times \sqrt{3} = \sqrt{2 \times 3} = \sqrt{6}$.

3. $\sqrt{3675}$

$$\sqrt{3675} = \sqrt{3 \times 5^2 \times 7^2}$$
$$= \sqrt{3} \times \sqrt{5^2} \times \sqrt{7^2}$$
$$= 5 \times 7 \times \sqrt{3}$$
$$= 35\sqrt{3}$$

> Since $\sqrt{3}$ is not a perfect square, I will leave it as it is.

4. $\sqrt{121}$

$$\sqrt{121} = \sqrt{11^2}$$
$$= 11$$

5. What is the length of the unknown side of the right triangle? Simplify your answer, if possible.

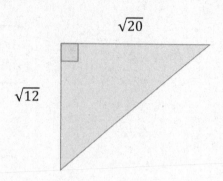

Let c units represent the length of the hypotenuse.

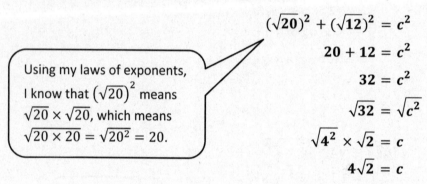

$$(\sqrt{20})^2 + (\sqrt{12})^2 = c^2$$
$$20 + 12 = c^2$$
$$32 = c^2$$
$$\sqrt{32} = \sqrt{c^2}$$
$$\sqrt{4^2} \times \sqrt{2} = c$$
$$4\sqrt{2} = c$$

Using my laws of exponents, I know that $\left(\sqrt{20}\right)^2$ means $\sqrt{20} \times \sqrt{20}$, which means $\sqrt{20 \times 20} = \sqrt{20^2} = 20$.

The length of the unknown side of the right triangle is $4\sqrt{2}$ units.

EUREKA
MATH

Simplify each of the square roots in Problems 1–5 as much as possible.

1. $\sqrt{98}$

2. $\sqrt{54}$

3. $\sqrt{144}$

4. $\sqrt{512}$

5. $\sqrt{756}$

6. What is the length of the unknown side of the right triangle? Simplify your answer, if possible.

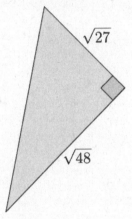

7. What is the length of the unknown side of the right triangle? Simplify your answer, if possible.

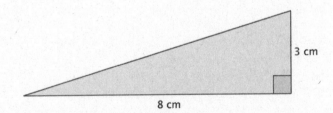

8. What is the length of the unknown side of the right triangle? Simplify your answer, if possible.

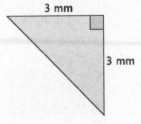

9. What is the length of the unknown side of the right triangle? Simplify your answer, if possible.

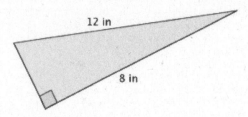

12 in

8 in

10. Josue simplified $\sqrt{450}$ as $15\sqrt{2}$. Is he correct? Explain why or why not.

11. Tiah was absent from school the day that you learned how to simplify a square root. Using $\sqrt{360}$, write Tiah an explanation for simplifying square roots.

EUREKA
MATH

Example 1

$$x^3 + 9x = \frac{1}{2}(18x + 54)$$

Example 2

$$x(x - 3) - 51 = -3x + 13$$

Exercises

Find the positive value of x that makes each equation true, and then verify your solution is correct.

1.

   a.   Solve $x^2 - 14 = 5x + 67 - 5x$.

   b.   Explain how you solved the equation.

2.   Solve and simplify: $x(x - 1) = 121 - x$

3.   A square has a side length of $3x$ inches and an area of 324 in². What is the value of x?

Lesson 5:   Solving Equations with Radicals

EUREKA
MATH

4. $-3x^3 + 14 = -67$

5. $x(x + 4) - 3 = 4(x + 19.5)$

6. $216 + x = x(x^2 - 5) + 6x$

EUREKA
MATH®

7.

    a.   What are we trying to determine in the diagram below?

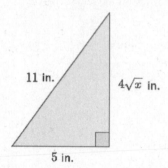

11 in.

$4\sqrt{x}$ in.

5 in.

    b.   Determine the value of x, and check your answer.

         Lesson 5:   Solving Equations with Radicals

EUREKA
MATH

Lesson Summary

Equations that contain variables that are squared or cubed can be solved using the properties of equality and the definition of square and cube roots.

Simplify an equation until it is in the form of $x^2 = p$ or $x^3 = p$, where p is a positive rational number; then, take the square or cube root to determine the positive value of x.

Example:

Solve for x.

$$\frac{1}{2}(2x^2 + 10) = 30$$
$$x^2 + 5 = 30$$
$$x^2 + 5 - 5 = 30 - 5$$
$$x^2 = 25$$
$$\sqrt{x^2} = \sqrt{25}$$
$$x = 5$$

Check:

$$\frac{1}{2}(2(5)^2 + 10) = 30$$
$$\frac{1}{2}(2(25) + 10) = 30$$
$$\frac{1}{2}(50 + 10) = 30$$
$$\frac{1}{2}(60) = 30$$
$$30 = 30$$

Name _____   Date _____

1. Find the positive value of x that makes the equation true, and then verify your solution is correct.

$$x^2 + 4x = 4(x + 16)$$

2. Find the positive value of x that makes the equation true, and then verify your solution is correct.

$$(4x)^3 = 1728$$

EUREKA
MATH®

1. Find the positive value of x that makes each equation true, and then verify that your solution is correct.
$$x^2 + 6x - 12 = 6(x + 4)$$

> Using my properties of equalities and the distributive property, I want to rewrite the equation in the form of $x^2 = p$, where p is a positive rational number, and then take the square root and determine the value of x.

$$x^2 + 6x - 12 = 6(x + 4)$$
$$x^2 + 6x - 12 = 6x + 24$$
$$x^2 + 6x - 12 + 12 = 6x + 24 + 12$$
$$x^2 + 6x = 6x + 36$$
$$x^2 + 6x - 6x = 6x - 6x + 36$$
$$x^2 = 36$$
$$x = 6$$

Check
$$6^2 + 6(6) - 12 = 6(6 + 4)$$
$$36 + 36 - 12 = 6(10)$$
$$60 = 60$$

2. Determine the positive value of x that makes the equation true, and then explain how you solved the equation.
$$\frac{x^{11}}{x^8} - 125 = 0$$

> I can rewrite the equation using the laws of exponents.

$$\frac{x^{11}}{x^8} - 125 = 0$$
$$x^3 - 125 = 0$$
$$x^3 - 125 + 125 = 0 + 125$$
$$x^3 = 125$$
$$\sqrt[3]{x^3} = \sqrt[3]{125}$$
$$x = \sqrt[3]{5^3}$$
$$x = 5$$

Check:
$$\frac{5^{11}}{5^8} - 125 = 0$$
$$5^3 - 125 = 0$$
$$125 - 125 = 0$$
$$0 = 0$$

To solve the equation, I first had to simplify the expression $\frac{x^{11}}{x^8}$ to x^3. Next, I used the properties of equality to transform the equation into $x^3 = 125$. Finally, I had to take the cube root of both sides of the equation to solve for x.

EUREKA MATH®

3. Determine the positive value of x that makes the equation true.

$$(5\sqrt{x})^2 - 7x = 72$$

> I only need to find the positive value of x that makes the equation true.

$$(5\sqrt{x})^2 - 7x = 72$$
$$5^2(\sqrt{x})^2 - 7x = 72$$
$$25x - 7x = 72$$
$$18x = 72$$
$$\frac{18x}{18} = \frac{72}{18}$$
$$x = 4$$

Check:

$$(5\sqrt{4})^2 - 7(4) = 72$$
$$5^2(\sqrt{4})^2 - 28 = 72$$
$$25(4) - 28 = 72$$
$$100 - 28 = 72$$
$$72 = 72$$

4. Determine the length of the hypotenuse of the right triangle below.

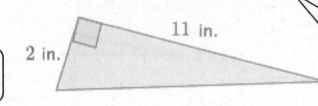

> The lengths of the sides of a triangle are positive, so a negative answer would not make sense.

> I must define my variable.

Let x in. represent the length of the hypotenuse.

$$2^2 + 11^2 = x^2$$
$$4 + 121 = x^2$$
$$125 = x^2$$
$$\sqrt{125} = \sqrt{x^2}$$
$$\sqrt{5^2 \cdot 5} = x$$
$$5\sqrt{5} = x$$

Check:

$$2^2 + 11^2 = (5\sqrt{5})^2$$
$$4 + 121 = 5^2(\sqrt{5})^2$$
$$125 = 25(5)$$
$$125 = 125$$

> I am using the laws of exponents to check my answer.

Since $x = 5\sqrt{5}$, then the length of the hypotenuse is $5\sqrt{5}$ in.

> This is an exact answer and cannot be simplified any further.

EUREKA MATH®

Find the positive value of x that makes each equation true, and then verify your solution is correct.

1. $x^2(x + 7) = \frac{1}{2}\left(14x^2 + 16\right)$

2. $x^3 = 1331^{-1}$

3. Determine the positive value of x that makes the equation true, and then explain how you solved the equation.

$$\frac{x^9}{x^7} - 49 = 0$$

4. Determine the positive value of x that makes the equation true.

$$(8x)^2 = 1$$

5. $\left(9\sqrt{x}\right)^2 - 43x = 76$

6. Determine the length of the hypotenuse of the right triangle below.

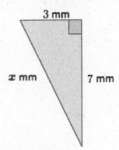

7. Determine the length of the legs in the right triangle below.

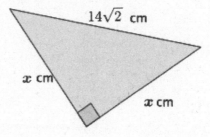

8. An equilateral triangle has side lengths of 6 cm. What is the height of the triangle? What is the area of the triangle?

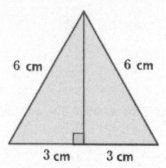

6 cm   6 cm

3 cm   3 cm

9. Challenge: Find the positive value of x that makes the equation true.

$$\left(\frac{1}{2}x\right)^2 - 3x = 7x + 8 - 10x$$

10. Challenge: Find the positive value of x that makes the equation true.

$$11x + x(x - 4) = 7(x + 9)$$

EUREKA
MATH

Opening Exercise

a. Use long division to determine the decimal expansion of $\frac{54}{20}$.

b. Use long division to determine the decimal expansion of $\frac{7}{8}$.

c. Use long division to determine the decimal expansion of $\frac{8}{9}$.

d. Use long division to determine the decimal expansion of $\frac{22}{7}$.

e. What do you notice about the decimal expansions of parts (a) and (b) compared to the decimal expansions of parts (c) and (d)?

Example 1

Consider the fraction $\frac{5}{8}$. Write an equivalent form of this fraction with a denominator that is a power of 10, and write the decimal expansion of this fraction.

Example 2

Consider the fraction $\frac{17}{125}$. Is it equal to a finite or an infinite decimal? How do you know?

EUREKA
MATH

Exercises 1–5

You may use a calculator, but show your steps for each problem.

1. Consider the fraction $\frac{3}{8}$.

   a. Write the denominator as a product of 2's and/or 5's. Explain why this way of rewriting the denominator helps to find the decimal representation of $\frac{3}{8}$.

   b. Find the decimal representation of $\frac{3}{8}$. Explain why your answer is reasonable.

2. Find the first four places of the decimal expansion of the fraction $\frac{43}{64}$.

3. Find the first four places of the decimal expansion of the fraction $\frac{29}{125}$.

4. Find the first four decimal places of the decimal expansion of the fraction $\frac{19}{34}$.

5. Identify the type of decimal expansion for each of the numbers in Exercises 1–4 as finite or infinite. Explain why their decimal expansion is such.

Example 3

Will the decimal expansion of $\frac{7}{80}$ be finite or infinite? If it is finite, find it.

Example 4

Will the decimal expansion of $\frac{3}{160}$ be finite or infinite? If it is finite, find it.

Lesson 6:   Finite and Infinite Decimals

Exercises 6–8

You may use a calculator, but show your steps for each problem.

6. Convert the fraction $\frac{37}{40}$ to a decimal.

    a. Write the denominator as a product of 2's and/or 5's. Explain why this way of rewriting the denominator helps to find the decimal representation of $\frac{37}{40}$.

    b. Find the decimal representation of $\frac{37}{40}$. Explain why your answer is reasonable.

7. Convert the fraction $\frac{3}{250}$ to a decimal.

8. Convert the fraction $\frac{7}{1250}$ to a decimal.

Lesson Summary

Fractions with denominators that can be expressed as products of 2's and/or 5's are equivalent to fractions with denominators that are a power of 10. These are precisely the fractions with finite decimal expansions.

Example:

Does the fraction $\frac{1}{8}$ have a finite or an infinite decimal expansion?

Since $8 = 2^3$, then the fraction has a finite decimal expansion. The decimal expansion is found as

$$\frac{1}{8} = \frac{1}{2^3} = \frac{1 \times 5^3}{2^3 \times 5^3} = \frac{125}{10^3} = 0.125.$$

If the denominator of a (simplified) fraction cannot be expressed as a product of 2's and/or 5's, then the decimal expansion of the number will be infinite.

EUREKA
MATH

Name _____   Date _____

Convert each fraction to a finite decimal if possible. If the fraction cannot be written as a finite decimal, then state how you know. You may use a calculator, but show your steps for each problem.

1. $\dfrac{9}{16}$

2. $\dfrac{8}{125}$

3. $\dfrac{4}{15}$

4. $\dfrac{1}{200}$

Convert each fraction to a finite decimal. If the fraction cannot be written as a finite decimal, then state how you know. Show your steps, but use a calculator for the multiplication.

1. $\dfrac{13}{16}$

    The denominator 16 is equal to 2^4.

    $$\frac{13}{16} = \frac{13}{2^4} = \frac{13 \times 5^4}{2^4 \times 5^4} = \frac{13 \times 5^4}{(2 \times 5)^4} = \frac{8125}{10^4} = 0.8125$$

    > The denominator 16 is equal to 2^4. I need to multiply the denominator by 4 factors of 5 so that the denominator is a multiple of 10, and I can write the fraction easily.

2. $\dfrac{2}{125}$

    The denominator 125 is equal to 5^3.

    $$\frac{2}{125} = \frac{2}{5^3} = \frac{2 \times 2^3}{2^3 \times 5^3} = \frac{2^4}{(2 \times 5)^3} = \frac{16}{10^3} = 0.016$$

    > I need to multiply the denominator by 3 factors of 2.

3. $\dfrac{22}{75}$

    The fraction $\dfrac{22}{75}$ is not a finite decimal because the denominator 75 is equal to 3×5^2. The denominator cannot be expressed as a product of 2's and 5's; therefore, $\dfrac{22}{75}$ is not a finite decimal.

    > The denominator cannot be written as a product of 2's and/or 5's.

4. $\dfrac{33}{800}$

    The denominator 800 is equal to $2^5 \times 5^2$.

    $$\frac{33}{800} = \frac{33}{2^5 \times 5^2} = \frac{33 \times 5^3}{2^5 \times 5^2 \times 5^3} = \frac{33 \times 5^3}{2^5 \times 5^5} = \frac{33 \times 5^3}{(2 \times 5)^5} = \frac{4125}{10^5} = 0.04125$$

    > To make the powers of the bases match, I need to multiply 5^2 by 5^3.

Convert each fraction given to a finite decimal, if possible. If the fraction cannot be written as a finite decimal, then state how you know. You may use a calculator, but show your steps for each problem.

1. $\dfrac{2}{32}$

2. $\dfrac{99}{125}$

3. $\dfrac{15}{128}$

4. $\dfrac{8}{15}$

5. $\dfrac{3}{28}$

6. $\dfrac{13}{400}$

7. $\dfrac{5}{64}$

8. $\dfrac{15}{35}$

9. $\dfrac{199}{250}$

10. $\dfrac{219}{625}$

Opening Exercise

a. Write the expanded form of the decimal 0.3765 using powers of 10.

b. Write the expanded form of the decimal 0.3333333… using powers of 10.

c. Have you ever wondered about the value of 0.99999…? Some people say this infinite decimal has value 1. Some disagree. What do you think?

Example 1

The number 0.253 is represented on the number line below.

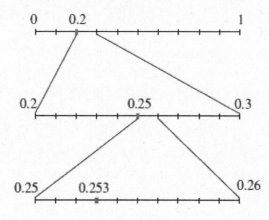

Example 2

The number $\frac{5}{6}$ which is equal to 0.833333… or $0.8\overline{3}$ is partially represented on the number line below.

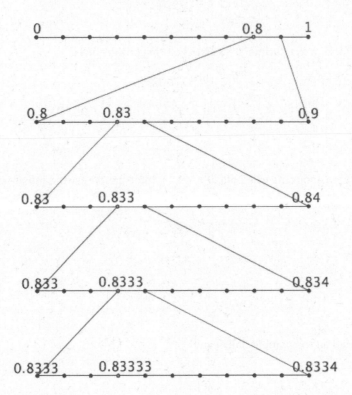

Lesson 7:   Infinite Decimals

© 2019 Great Minds®. eureka-math.org

EUREKA
MATH®

Exercises 1–5

1.

    a. Write the expanded form of the decimal 0.125 using powers of 10.

    b. Show on the number line the placement of the decimal 0.125.

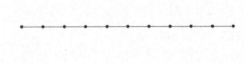

2.

    a. Write the expanded form of the decimal 0.3875 using powers of 10.

    b. Show on the number line the placement of the decimal 0.3875.

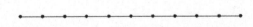

3.

a. Write the expanded form of the decimal 0.777777... using powers of 10.

b. Show the first few stages of placing the decimal 0.777777... on the number line.

EUREKA
MATH

4.

a. Write the expanded form of the decimal $0.\overline{45}$ using powers of 10.

b. Show the first few stages of placing the decimal $0.\overline{45}$ on the number line.

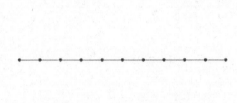

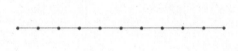

5.

a. Order the following numbers from least to greatest: 2.121212, 2.1, 2.2, and $2.\overline{12}$.

b. Explain how you knew which order to put the numbers in.

EUREKA MATH

Lesson Summary

An infinite decimal is a decimal whose expanded form is infinite.

Example:

The expanded form of the decimal $0.8\overline{3} = 0.83333\ldots$ is $\frac{8}{10} + \frac{3}{10^2} + \frac{3}{10^3} + \frac{3}{10^4} + \cdots$.

To pin down the placement of an infinite decimal on the number line, we first identify within which tenth it lies, then within which hundredth it lies, then within which thousandth, and so on. These intervals have widths getting closer and closer to a width of zero.

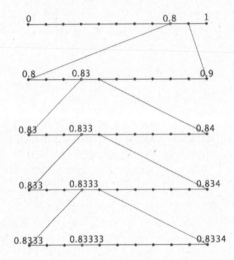

This reasoning allows us to deduce that the infinite decimal 0.9999... and 1 have the same location on the number line. Consequently, $0.\overline{9} = 1$.

EUREKA
MATH

Name _____   Date _____

1.

    a.   Write the expanded form of the decimal 0.829 using powers of 10.

    b.   Show the placement of the decimal 0.829 on the number line.

0   1

2.

    a.   Write the expanded form of the decimal 0.55555… using powers of 10.

    b.   Show the first few stages of placing the decimal 0.555555… on the number line.

Lesson 7:   Infinite Decimals

EUREKA MATH

3.

    a. Write the expanded form of the decimal $0.\overline{573}$ using powers of 10.

    b. Show the first few stages of placing the decimal $0.\overline{573}$ on the number line.

1.

a. Write the expanded form of the decimal 0.532 using powers of 10.

$$0.532 = \frac{5}{10} + \frac{3}{10^2} + \frac{2}{10^3}$$

> The first number line is divided into 10 equal parts, namely tenths. The next number line is divided into 10 equal parts, namely hundredths. The third number line is divided into 10 equal parts, namely thousandths.

b. Show on the number line the representation of the decimal 0.532.

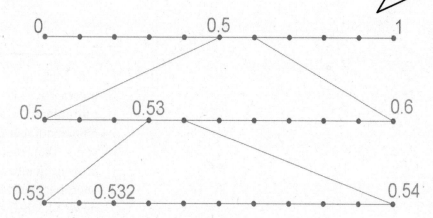

c. Is the decimal finite or infinite? How do you know?

The decimal 0.532 is finite because it can be completely represented by a finite number of steps in the sequence.

2.

a. Write the expanded form of the decimal $0.\overline{14}$ using powers of 10.

$$0.\overline{14} = \frac{1}{10} + \frac{4}{10^2} + \frac{1}{10^3} + \frac{4}{10^4} + \frac{1}{10^5} + \frac{4}{10^6} + \dots$$

b. Show on the number line the representation of the decimal 0.141414....

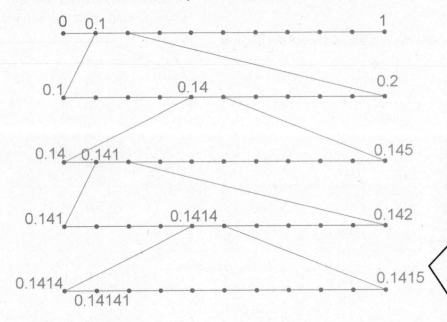

I see that the number $0.\overline{14}$ is represented by an infinite number of steps. I can keep drawing number lines, but I will never find the precise location of $0.\overline{14}$.

c. Is the decimal finite or infinite? How do you know?

The decimal $0.\overline{14}$ is infinite because it cannot be represented by a finite number of steps. Because the digits 1 and 4 continue to repeat, there will be an infinite number of steps in the sequence.

3. Explain why $0.444 < 0.4444$.

Rewriting each decimal in its expanded form will help me to explain which is larger.

The number $0.444 = \frac{4}{10} + \frac{4}{10^2} + \frac{4}{10^3}$, and the number

$0.4444 = \frac{4}{10} + \frac{4}{10^2} + \frac{4}{10^3} + \frac{4}{10^4}$. *That means that 0.4444*

is exactly $\frac{4}{10^4}$ larger than 0.444. If we examine the numbers

on the number line, 0.4444 is to the right of 0.444,

meaning that it is larger than 0.444.

1.

   a. Write the expanded form of the decimal 0.625 using
       powers of 10.

   b. Place the decimal 0.625 on the number line.

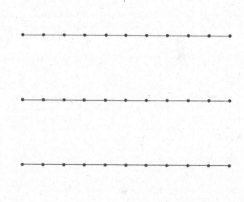

2.

   a. Write the expanded form of the decimal $0.\overline{370}$ using
       powers of 10.

   b. Show the first few stages of placing the decimal 0.370370... on
       the number line.

3. Which is a more accurate representation of the fraction $\frac{2}{3}$: 0.6666 or $0.\overline{6}$? Explain. Which would you prefer to
    compute with?

4. Explain why we shorten infinite decimals to finite decimals to perform operations. Explain the effect of shortening
    an infinite decimal on our answers.

5. A classmate missed the discussion about why $0.\overline{9} = 1$. Convince your classmate that this equality is true.

6. Explain why $0.3333 < 0.33333$.

Example 1

Show that the decimal expansion of $\frac{26}{4}$ is 6.5.

Exploratory Challenge/Exercises 1–5

1.

     a.   Use long division to determine the decimal expansion of $\frac{142}{2}$.

     b.   Fill in the blanks to show another way to determine the decimal expansion of $\frac{142}{2}$.

$$142 = \underline{\quad} \times 2 + \underline{\quad}$$

$$\frac{142}{2} = \frac{\underline{\quad} \times 2 + \underline{\quad}}{2}$$

$$\frac{142}{2} = \frac{\underline{\quad} \times 2}{2} + \frac{\underline{\quad}}{2}$$

$$\frac{142}{2} = \underline{\quad} + \frac{\underline{\quad}}{2}$$

$$\frac{142}{2} = \underline{\quad}$$

   c.   Does the number $\dfrac{142}{2}$ have a finite or an infinite decimal expansion?

2.

   a.   Use long division to determine the decimal expansion of $\dfrac{142}{4}$.

   b.   Fill in the blanks to show another way to determine the decimal expansion of $\dfrac{142}{4}$.

$$142 = \underline{\quad} \times 4 + \underline{\quad}$$

$$\frac{142}{4} = \frac{\underline{\quad} \times 4 + \underline{\quad}}{4}$$

$$\frac{142}{4} = \frac{\underline{\quad} \times 4}{4} + \frac{\underline{\quad}}{4}$$

$$\frac{142}{4} = \underline{\quad} + \frac{\underline{\quad}}{4}$$

$$\frac{142}{4} = \underline{\quad\quad\quad}$$

   c.   Does the number $\dfrac{142}{4}$ have a finite or an infinite decimal expansion?

     Lesson 8:   The Long Division Algorithm

EUREKA MATH®

3.

   a.   Use long division to determine the decimal expansion of $\frac{142}{6}$.

   b.   Fill in the blanks to show another way to determine the decimal expansion of $\frac{142}{6}$.

$$142 \;=\; \underline{\quad} \times 6 + \underline{\quad}$$

$$\frac{142}{6} \;=\; \frac{\underline{\quad} \times 6 + \underline{\quad}}{6}$$

$$\frac{142}{6} \;=\; \frac{\underline{\quad} \times 6}{6} + \frac{\underline{\quad}}{6}$$

$$\frac{142}{6} \;=\; \underline{\quad} + \frac{\underline{\quad}}{6}$$

$$\frac{142}{6} \;=\; \underline{\qquad\qquad}$$

   c.   Does the number $\frac{142}{6}$ have a finite or an infinite decimal expansion?

4.

   a. Use long division to determine the decimal expansion of $\frac{142}{11}$.

   b. Fill in the blanks to show another way to determine the decimal expansion of $\frac{142}{11}$.

$$142 = \underline{\quad} \times 11 + \underline{\quad}$$

$$\frac{142}{11} = \frac{\underline{\quad} \times 11 + \underline{\quad}}{11}$$

$$\frac{142}{11} = \frac{\underline{\quad} \times 11}{11} + \frac{\underline{\quad}}{11}$$

$$\frac{142}{11} = \underline{\quad} + \frac{\underline{\quad}}{11}$$

$$\frac{142}{11} = \underline{\qquad\qquad}$$

   c. Does the number $\frac{142}{11}$ have a finite or an infinite decimal expansion?

Lesson 8:   The Long Division Algorithm

EUREKA
MATH

5. In general, which fractions produce infinite decimal expansions?

Exercises 6–10

6. Does the number $\frac{65}{13}$ have a finite or an infinite decimal expansion? Does its decimal expansion have a repeating pattern?

7. Does the number $\frac{17}{11}$ have a finite or an infinite decimal expansion? Does its decimal expansion have a repeating pattern?

8. Is the number 0.212112111211112111112... rational? Explain. (Assume the pattern you see in the decimal expansion continues.)

9. Does the number $\frac{860}{999}$ have a finite or an infinite decimal expansion? Does its decimal expansion have a repeating pattern?

10. Is the number 0.123456789101112131415161718192021222... rational? Explain. (Assume the pattern you see in the decimal expansion continues.)

EUREKA
MATH

Lesson Summary

A rational number is a number that can be written in the form $\frac{a}{b}$ for a pair of integers a and b with b not zero.

The long division algorithm shows that every rational number has a decimal expansion that falls into a repeating pattern. For example, the rational number 32 has a decimal expansion of $32.\overline{0}$, the rational number $\frac{1}{3}$ has a decimal expansion of $0.\overline{3}$, and the rational number $\frac{4}{11}$ has a decimal expansion of $0.\overline{36}$.

© 2019 Great Minds®. eureka-math.org

Name _____ Date _____

1. Will the decimal expansion of $\frac{125}{8}$ be finite or infinite? Explain. If we were to write the decimal expansion of this rational number as an infinitely long decimal, which block of numbers repeat?

2. Write the decimal expansion of $\frac{13}{7}$ as an infinitely long repeating decimal.

1. Write the decimal expansion of $\frac{4000}{3}$. Based on our definition of rational numbers having a decimal expansion that repeats eventually, is the number rational? Explain.

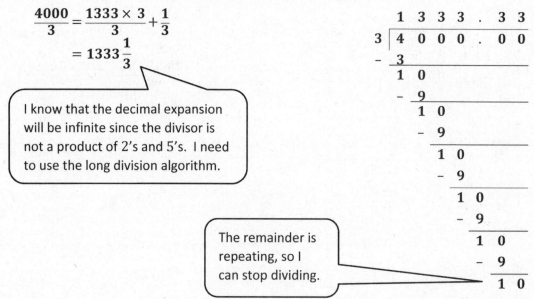

$$\frac{4000}{3} = \frac{1333 \times 3}{3} + \frac{1}{3}$$

$$= 1333\frac{1}{3}$$

I know that the decimal expansion will be infinite since the divisor is not a product of 2's and 5's. I need to use the long division algorithm.

The remainder is repeating, so I can stop dividing.

The decimal expansion of $\frac{4000}{3}$ is $1333.\overline{3}$. The number is rational because it has the repeating digit of 3. Rational numbers have decimal expansions that repeat; therefore, $\frac{4000}{3}$ is a rational number.

2. Write the decimal expansion of $\frac{3,888,885}{11}$. Based on our definition of rational numbers having a decimal expansion that repeats eventually, is the number rational? Explain.

$$\frac{3\,888\,885}{11} = \frac{353\,535 \times 11}{11} + \frac{0}{11}$$

$$= 353\,535$$

The $+\frac{0}{11}$ means that the remainder is a repeating block of zeros.

The decimal expansion of $\frac{3\,888\,885}{11}$ is $353,535$. The number is rational because we can write the repeating digit of 0 following the whole number. Rational numbers have decimal expansions that repeat; therefore, $\frac{3\,888\,885}{11}$ is a rational number.

3. Someone notices that the long division of 1,855,855 by 9 has a quotient of 206,206 and a remainder of 1 and wonders why there is a repeating block of digits in the quotient, namely 206. Explain to the person why this happens.

$$
\require{enclose}
\begin{array}{r}
2\ 0\ 6\ 2\ 0\ 6 \\
9 \enclose{longdiv}{1\ 8\ 5\ 5\ 8\ 5\ 5} \\
\end{array}
$$

	2	0	6	2	0	6
9	1	8	5	5	8	5	5
−	1	8				
	0	5				
−		0				
		5	5			
−		5	4			
		1	8			
−		1	8			
		0	5			
−			0			
			5	5		
−			5	4		
				1		

$$\frac{1\,855\,855}{9} = \frac{206\,206 \times 9}{9} + \frac{1}{9}$$

$$= 206\,206\frac{1}{9}$$

The block of digits 206 repeats because the long division algorithm leads us to perform the same division over and over again. In the algorithm shown above, we see that there are two groups of 9 in 18, leaving a remainder of 0. When we bring down the 5, we see that there are zero groups of 9 in 5. When we bring down the next 5, we see that there are six groups of 9 in 55, leaving a remainder of 1. That is when the process starts over because the next step is to bring down 8, giving us 18, which is what we started with. Since the division repeats, then the digits in the quotient will repeat.

4. Is the number $\frac{15}{11} = 1.36363636\ldots$ rational? Explain.

The number appears to be rational because the decimal expansion has a repeat block of 36. Because every rational number has a block that repeats, the number is rational.

> I can rewrite $\frac{15}{11} = 1.\overline{36}$ because the decimal expansion has a repeating block of 36.

5. Is the number $\sqrt{5} = 2.23606787\ldots$ rational? Explain.

The number appears to have a decimal expansion that does not have decimal digits that repeat in a block. For that reason, this is not a rational number.

> I don't see any pattern to the decimal expansion.

Lesson 8: The Long Division Algorithm

EUREKA MATH®

1. Write the decimal expansion of $\dfrac{7000}{9}$ as an infinitely long repeating decimal.

2. Write the decimal expansion of $\dfrac{6555555}{3}$ as an infinitely long repeating decimal.

3. Write the decimal expansion of $\dfrac{350000}{11}$ as an infinitely long repeating decimal.

4. Write the decimal expansion of $\dfrac{12,000,000}{37}$ as an infinitely long repeating decimal.

5. Someone notices that the long division of 2,222,222 by 6 has a quotient of 370,370 and a remainder of 2 and wonders why there is a repeating block of digits in the quotient, namely 370. Explain to the person why this happens.

6. Is the answer to the division problem number 10 ÷ 3.2 a rational number? Explain.

7. Is $\dfrac{3\pi}{77\pi}$ a rational number? Explain.

8. The decimal expansion of a real number x has every digit 0 except the first digit, the tenth digit, the hundredth digit, the thousandth digit, and so on, are each 1. Is x a rational number? Explain.

Opening Exercise

a. Compute the decimal expansions of $\frac{5}{6}$ and $\frac{7}{9}$.

b. What is $\frac{5}{6} + \frac{7}{9}$ as a fraction? What is the decimal expansion of this fraction?

c. What is $\frac{5}{6} \times \frac{7}{9}$ as a fraction? According to a calculator, what is the decimal expansion of the answer?

d. If you were given just the decimal expansions of $\frac{5}{6}$ and $\frac{7}{9}$, without knowing which fractions produced them, do you think you could easily add the two decimals to find the decimal expansion of their sum? Could you easily multiply the two decimals to find the decimal expansion of their product?

Exercise 1

Two irrational numbers x and y have infinite decimal expansions that begin $0.67035267\ldots$ for x and $0.84991341\ldots$ for y.

a. Explain why 0.670 is an approximation for x with an error of less than one thousandth. Explain why 0.849 is an approximation for y with an error of less than one thousandth.

b. Using the approximations given in part (a), what is an approximate value for $x + y$, for $x \times y$, and for $x^2 + 7y^2$?

c. Repeat part (b), but use approximations for x and y that have errors less than $\dfrac{1}{10^5}$.

EUREKA
MATH

Exercise 2

Two real numbers have decimal expansions that begin with the following:

$$x = 0.1538461....$$
$$y = 0.3076923....$$

a. Using approximations for x and y that are accurate within a measure of $\frac{1}{10^3}$, find approximate values for $x + y$ and $y - 2x$.

b. Using approximations for x and y that are accurate within a measure of $\frac{1}{10^7}$, find approximate values for $x + y$ and $y - 2x$.

c. We now reveal that $x = \frac{2}{13}$ and $y = \frac{4}{13}$. How accurate is your approximate value to $y - 2x$ from part (a)? From part (b)?

d. Compute the first seven decimal places of $\frac{6}{13}$. How accurate is your approximate value to $x + y$ from part (a)? From part (b)?

> **Lesson Summary**
>
> It is not clear how to perform arithmetic on numbers given as infinitely long decimals. If we approximate these numbers by truncating their infinitely long decimal expansions to a finite number of decimal places, then we can perform arithmetic on the approximate values to estimate answers.
>
> Truncating a decimal expansion to n decimal places gives an approximation with an error of less than $\dfrac{1}{10^n}$. For example, 0.676 is an approximation for 0.676767… with an error of less than 0.001.

EUREKA
MATH®

Name _____   Date _____

Suppose $x = \dfrac{2}{3} = 0.6666\ldots$ and $y = \dfrac{5}{9} = 0.5555\ldots$

a. Using 0.666 as an approximation for x and 0.555 as an approximation for y, find an approximate value for $x + y$.

b. What is the true value of $x + y$ as an infinite decimal?

c. Use approximations for x and y, each accurate to within an error of $\dfrac{1}{10^5}$, to estimate a value of the product $x \times y$.

1. Choose a power of 10 to convert this fraction to a decimal: $\dfrac{3}{11}$.
   Choices will vary.

> It is better to use a higher power of 10; extra zeros will not change the value of the decimal expansion.

The work shown below uses the factor 10^6. Students should choose a factor of at least 10^4 in order to get an approximate decimal expansion and notice that the decimal expansion repeats.

a. Determine the decimal expansion of $\dfrac{3}{11}$. Verify you are correct using a calculator.

$$\dfrac{3}{11} = \dfrac{3 \times 10^6}{11} \times \dfrac{1}{10^6}$$
$$= \dfrac{3\,000\,000}{11} \times \dfrac{1}{10^6}$$

$$3\,000\,000 = 272\,727 \times 11 + 3$$

> I can use division with remainder to find that $3{,}000{,}000 = 272{,}727 \times 11 + 3$.

$$\dfrac{3}{11} = \dfrac{272\,727 \times 11 + 3}{11} \times \dfrac{1}{10^6}$$
$$= \left(\dfrac{272\,727 \times 11}{11} + \dfrac{3}{11}\right) \times \dfrac{1}{10^6}$$
$$= \left(272\,727 + \dfrac{3}{11}\right) \times \dfrac{1}{10^6}$$
$$= 272\,727 \times \dfrac{1}{10^6} + \left(\dfrac{3}{11} \times \dfrac{1}{10^6}\right)$$
$$= \dfrac{272\,727}{10^6} + \left(\dfrac{3}{11} \times \dfrac{1}{10^6}\right)$$
$$= 0.272\,727 + \left(\dfrac{3}{11} \times \dfrac{1}{10^6}\right)$$

The decimal expansion of $\dfrac{3}{11}$ is approximately 0.272727.

2. Ted wrote the decimal expansion of $\frac{5}{7}$ as 0.724285, but when he checked it on a calculator, it was 0.714285. Identify the error from his work below and explain what he did wrong.

$$\frac{5}{7} = \frac{5 \times 10^6}{7} \times \frac{1}{10^6}$$

$$= \frac{5\,000\,000}{7} \times \frac{1}{10^6}$$

$$5\,000\,000 = 724\,285 \times 7 + 5$$

$$\frac{5}{7} = \frac{724\,285 \times 7 + 5}{7} \times \frac{1}{10^6}$$

$$= \left(\frac{724\,285 \times 7}{7} + \frac{5}{7}\right) \times \frac{1}{10^6}$$

$$= \left(724\,285 + \frac{5}{7}\right) \times \frac{1}{10^6}$$

$$= 724\,285 \times \frac{1}{10^6} + \left(\frac{5}{7} \times \frac{1}{10^6}\right)$$

$$= \frac{724\,285}{10^6} + \left(\frac{5}{7} \times \frac{1}{10^6}\right)$$

$$= 0.724285 + \left(\frac{5}{7} \times \frac{1}{10^6}\right)$$

Ted did the division with remainder incorrectly. He wrote that $5,000,000 = 724,285 \times 7 + 5$ when actually $5,000,000 = 714,285 \times 7 + 5$. This error led to his decimal expansion being incorrect.

> This is such a small value that it won't affect the estimate. In fact, $\left(\frac{4}{7} \times \frac{1}{10^6}\right) < 0.000001$.

3. Given that $\frac{4}{7} = 0.571428 + \left(\frac{4}{7} \times \frac{1}{10^6}\right)$, explain why 0.571248 is a good estimate of $\frac{4}{7}$.

When you consider the value of $\left(\frac{4}{7} \times \frac{1}{10^6}\right)$, then it is clear that 0.571428 is a good estimate of $\frac{4}{7}$. We know that $\frac{4}{7} < 1$. By the basic inequality property, we also know that $\frac{4}{7} \times \frac{1}{10^6} < 1 \times \frac{1}{10^6}$, which means that $\frac{4}{7} \times \frac{1}{10^6} < 0.000001$. That is such a small value that it will not affect the estimate of $\frac{4}{7}$ in any real way.

1. Two irrational numbers x and y have infinite decimal expansions that begin $0.3338117\ldots$ for x and $0.9769112\ldots$ for y.

   a. Explain why 0.33 is an approximation for x with an error of less than one hundredth. Explain why 0.97 is an approximation for y with an error of less than one hundredth.

   b. Using the approximations given in part (a), what is an approximate value for $2x(y+1)$?

   c. Repeat part (b), but use approximations for x and y that have errors less than $\dfrac{1}{10^6}$.

2. Two real numbers have decimal expansions that begin with the following:

$$x = 0.70588\ldots.$$
$$y = 0.23529\ldots.$$

   a. Using approximations for x and y that are accurate within a measure of $\dfrac{1}{10^2}$, find approximate values for $x + 1.25y$ and $\dfrac{x}{y}$.

   b. Using approximations for x and y that are accurate within a measure of $\dfrac{1}{10^4}$, find approximate values for $x + 1.25y$ and $\dfrac{x}{y}$.

   c. We now reveal that x and y are rational numbers with the property that each of the values $x + 1.25y$ and $\dfrac{x}{y}$ is a whole number. Make a guess as to what whole numbers these values are, and use your guesses to find what fractions x and y might be.

Example 1

There is a fraction with an infinite decimal expansion of $0.\overline{81}$. Find the fraction.

Exercises 1–2

1. There is a fraction with an infinite decimal expansion of $0.\overline{123}$. Let $x = 0.\overline{123}$.

    a. Explain why looking at $1000x$ helps us find the fractional representation of x.

© 2019 Great Minds®. eureka-math.org

b. What is x as a fraction?

c. Is your answer reasonable? Check your answer using a calculator.

2. There is a fraction with a decimal expansion of $0.\overline{4}$. Find the fraction, and check your answer using a calculator.

EUREKA
MATH

Example 2

Could it be that $2.13\overline{8}$ is also a fraction?

Exercises 3–4

3. Find the fraction equal to $1.6\overline{23}$. Check your answer using a calculator.

4. Find the fraction equal to $2.9\overline{60}$. Check your answer using a calculator.

    Lesson 10:   Converting Repeating Decimals to Fractions

EUREKA
MATH

Lesson Summary

Every decimal with a repeating pattern is a rational number, and we have the means to determine the fraction that has a given repeating decimal expansion.

Example: Find the fraction that is equal to the number $0.\overline{567}$.

Let x represent the infinite decimal $0.\overline{567}$.

$$x = 0.\overline{567}$$
$$10^3 x = 10^3 (0.\overline{567}) \qquad \text{Multiply by } 10^3 \text{ because there are 3 digits that repeat.}$$
$$1000x = 567.\overline{567} \qquad \text{Simplify}$$
$$1000x = 567 + 0.\overline{567} \qquad \text{By addition}$$
$$1000x = 567 + x \qquad \text{By substitution; } x = 0.\overline{567}$$
$$1000x - x = 567 + x - x \qquad \text{Subtraction property of equality}$$
$$999x = 567 \qquad \text{Simplify}$$
$$\frac{999}{999} x = \frac{567}{999} \qquad \text{Division property of equality}$$
$$x = \frac{567}{999} = \frac{63}{111} \qquad \text{Simplify}$$

This process may need to be used more than once when the repeating digits, as in numbers such as $1.2\overline{6}$, do not begin immediately after the decimal.

Irrational numbers are numbers that are not rational. They have infinite decimal expansions that do not repeat and they cannot be expressed as $\frac{p}{q}$ for integers p and q with $q \neq 0$.

Name _____ Date _____

1. Find the fraction equal to $0.\overline{534}$.

2. Find the fraction equal to $3.0\overline{15}$.

1. Let $x = 0.\overline{912}$. Explain why multiplying both sides of this equation by 10^3 will help us determine the fractional representation of x.

   When we multiply both sides of the equation by 10^3, on the right side we will have $912.912912\ldots$. This is helpful because we will be able to subtract the repeating decimal from both sides by subtracting x.

   > I want to multiply x by 10^n, where n is the number of digits in the repeating block, so that the remaining decimal digits are those that repeat.

   a. After multiplying both sides of the equation by 10^3, rewrite the resulting equation by making a substitution that will help determine the fractional value of x. Explain how you were able to make the substitution.

   $$x = 0.\overline{912}$$
   $$(10^3)x = (10^3)(0.\overline{912})$$
   $$1000x = 912.\overline{912}$$
   $$1000x = 912 + 0.912\,912\ldots$$
   $$1000x = 912 + x$$

   > When I write $912.\overline{912}$ as a sum, I see that $0.\overline{912} = x$, and I can use substitution.

   Since we let $x = 0.\overline{912}$, we can substitute the repeating decimal $0.912912\ldots$ with x.

   b. Solve the equation to determine the value of x.

   $$1000x - x = 912 + x - x$$
   $$999x = 912$$
   $$\frac{999x}{999} = \frac{912}{999}$$
   $$x = \frac{912}{999}$$

   > I can use properties of equality to solve for x.

   c. Is your answer reasonable? Check your answer using a calculator.

   Yes, my answer is reasonable and correct. It is reasonable because the denominator cannot be expressed as a product of 2's and 5's; therefore, I know that the fraction must represent an infinite decimal. Also, the number $0.\overline{912}$ is closer to 1 than 0.5, and the fraction is also closer to 1 than $\frac{1}{2}$. It is correct because the division of $\frac{912}{999}$ using the calculator is $0.912912912\ldots$.

2. Find the fraction equal to $6.7\overline{56}$. Check your answer using a calculator.

> I need to let y equal the repeating block of my decimal. I will repeat the process I learned in class with both x and y.

Let $x = 6.7\overline{56}$.

$x = 6.7\overline{56}$

$10x = (10)6.7\overline{56}$

$10x = 67.\overline{56}$

> I am going to multiply by 10 so that the only remaining decimal digits are those that repeat.

Let $y = 0.\overline{56}$.

$y = 0.\overline{56}$

$100y = 100(0.\overline{56})$

$100y = 56.\overline{56}$

$100y = 56 + y$

$100y - y = 56 + y - y$

$99y = 56$

$\dfrac{99y}{99} = \dfrac{56}{99}$

$y = \dfrac{56}{99}$

$10x = 67.\overline{56}$

$10x = 67 + y$

$10x = 67 + \dfrac{56}{99}$

$10x = \dfrac{67 \times 99}{99} + \dfrac{56}{99}$

$10x = \dfrac{67 \times 99 + 56}{99}$

$10x = \dfrac{6689}{99}$

$\left(\dfrac{1}{10}\right)10x = \left(\dfrac{1}{10}\right)\dfrac{6689}{99}$

$x = \dfrac{6689}{990}$

$$6.7\overline{56} = \dfrac{6689}{990}$$

Lesson 10:   Converting Repeating Decimals to Fractions

© 2019 Great Minds®. eureka-math.org

EUREKA MATH

1.

    a. Let $x = 0.\overline{631}$. Explain why multiplying both sides of this equation by 10^3 will help us determine the fractional representation of x.

    b. What fraction is x?

    c. Is your answer reasonable? Check your answer using a calculator.

2. Find the fraction equal to $3.40\overline{8}$. Check your answer using a calculator.

3. Find the fraction equal to $0.\overline{5923}$. Check your answer using a calculator.

4. Find the fraction equal to $2.3\overline{82}$. Check your answer using a calculator.

5. Find the fraction equal to $0.\overline{714285}$. Check your answer using a calculator.

6. Explain why an infinite decimal that is not a repeating decimal cannot be rational.

7. In a previous lesson, we were convinced that it is acceptable to write $0.\overline{9} = 1$. Use what you learned today to show that it is true.

8. Examine the following repeating infinite decimals and their fraction equivalents. What do you notice? Why do you think what you observed is true?

$$0.\overline{81} = \frac{81}{99} \qquad 0.\overline{4} = \frac{4}{9} \qquad 0.\overline{123} = \frac{123}{999} \qquad 0.\overline{60} = \frac{60}{99}$$

$$0.\overline{4311} = \frac{4311}{9999} \qquad 0.\overline{01} = \frac{1}{99} \qquad 0.\overline{3} = \frac{1}{3} = \frac{3}{9} \qquad 0.\overline{9} = 1.0$$

Opening Exercise

Place $\sqrt{28}$ on a number line. Make a guess as to the first few values of the decimal expansion of $\sqrt{28}$. Explain your reasoning.

Example 1

Consider the decimal expansion of $\sqrt{3}$.

Find the first two values of the decimal expansion using the following fact: If $c^2 < 3 < d^2$ for positive numbers c and d, then $c < \sqrt{3} < d$.

First Approximation:

Because $1 < 3 < 4$, we have $1 < \sqrt{3} < 2$.

Second approximation:

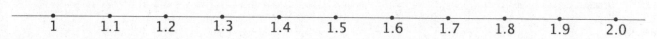

Third approximation:

Example 2

Find the first few places of the decimal expansion of $\sqrt{28}$.

First approximation:

Second approximation:

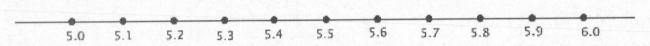

Third approximation:

Fourth approximation:

EUREKA
MATH

Exercise 1

In which interval of hundredths does $\sqrt{14}$ lie? Show your work.

Lesson Summary

To find the first few decimal places of the decimal expansion of the square root of a non-perfect square, first determine between which two integers the square root lies, then in which interval of a tenth the square root lies, then in which interval of a hundredth it lies, and so on.

Example: Find the first few decimal places of $\sqrt{22}$.

Begin by determining between which two integers the number would lie.

$\sqrt{22}$ is between the integers 4 and 5 because $16 < 22 < 25$.

Next, determine between which interval of tenths the number belongs.

$\sqrt{22}$ is between 4.6 and 4.7 because $4.6^2 = 21.16 < 22 < 4.7^2 = 22.09$.

Next, determine between which interval of hundredths the number belongs.

$\sqrt{22}$ is between 4.69 and 4.70 because $4.69^2 = 21.9961 < 22 < 4.70^2 = 22.0900$.

A good estimate of the value of $\sqrt{22}$ is 4.69. It is correct to two decimal places and so has an error no larger than 0.01.

Notice that with each step of this process we are getting closer and closer to the actual value $\sqrt{22}$. This process can continue using intervals of thousandths, ten-thousandths, and so on.

EUREKA
MATH

Name _____   Date _____

Determine the three-decimal digit approximation of the number $\sqrt{17}$.

1. Use the method of rational approximation to determine the decimal expansion of $\sqrt{42}$. Determine which interval of hundredths it would lie in.

    The number $\sqrt{42}$ is between 6 and 7 but closer to 6.

    Looking at the interval of tenths, beginning with 6.4 to 6.5, the number $\sqrt{42}$ lies between 6.4 and 6.5 because $6.4^2 = 40.96$ and $6.5^2 = 42.25$ but is closer to 6.5.

    In the interval of hundredths, the number $\sqrt{42}$ lies between 6.48 and 6.49 because $6.48^2 = 41.9904$ and $6.49^2 = 42.1201$.

    Since 42 is closer to 6.48^2 than 6.49^2, the number $\sqrt{42}$ is approximately 6.48.

    > In class, I learned rational approximation by examining the number at increasingly smaller powers such as tenths, hundredths, and so on.

    > I know $\sqrt{42}$ is between 6 and 7 because $6^2 < \left(\sqrt{42}\right)^2 < 7^2$. Since $\sqrt{42}$ is almost halfway between $\sqrt{36}$ and $\sqrt{49}$, I will start looking at 6.4.

2. Determine the three-decimal digit approximation of the number $\sqrt{83}$.

    The number $\sqrt{83}$ is between 9 and 10 but closer to 9.

    Looking at the interval of tenths, beginning with 9.1 to 9.2, the number $\sqrt{83}$ lies between 9.1 and 9.2 because $9.1^2 = 82.81$, and $9.2^2 = 84.64$ and is closer to 9.1.

    In the interval of hundredths, the number $\sqrt{83}$ lies between 9.11 and 9.12 because $9.11^2 = 82.9921$, and $9.12^2 = 83.1744$ and is closer to 9.11.

    In the interval of thousandths, the number $\sqrt{83}$ lies between 9.110 and 9.111 because $9.110^2 = 82.9921$, and $9.111^2 = 83.010321$ but is closer to 82.9921.

    Since 83 is closer to 9.110^2 than 9.111^2, the three-decimal digit approximation of the number is approximately 9.110.

    > So far my decimal approximation is $\sqrt{83} \approx 9.11$. But I need to continue this process for one more decimal place.

© 2019 Great Minds®. eureka-math.org

3. Is the number $\sqrt{360}$ rational or irrational? Explain your answer.

    The number $\sqrt{360}$ is an irrational number because it has a decimal expansion that is infinite and does not repeat. That is, the number $\sqrt{360}$ cannot be expressed as a rational number; therefore, it is irrational.

    > I learned that irrational numbers are infinite decimals that do not have repeating blocks of digits.

4. Is the number 0.94949494... rational or irrational?
    Explain your answer.

    > I learned how to convert repeating decimals to fractions in the previous lesson. Repeating blocks of digits means this is a rational number.

    The number 0.94949494... can be expressed as the fraction $\frac{94}{99}$; therefore, it is a rational number. Not only is the number $\frac{94}{99}$ a quotient of integers, but its decimal expansion is infinite with a repeating block of digits.

5. Challenge: Determine the two-decimal digit approximation of the number $\sqrt[3]{25}$.

    The number $\sqrt[3]{25}$ is between integers 2 and 3 because
    $$2^3 < \left(\sqrt[3]{25}\right)^3 < 3^3, \text{ or } 8 < 25 < 27.$$

    Since $\sqrt[3]{25}$ is closer to 3, I will start checking the tenths intervals between 2.9 and 3. $\sqrt[3]{25}$ is between 2.9 and 3 since $2.9^3 = 24.389$ and $3^3 = 27$.

    > When I use rational approximation, I need to cube the values instead of squaring them since I am trying to approximate the cube root of 25.

    Checking the hundredths interval, $\sqrt[3]{25}$ is between 2.92 and 2.93 since $2.92^3 = 24.897088$ and $2.93^3 = 25.153757$.

    Since 25 is closer to 2.92^3 than 2.93^3, the two-decimal digit approximation is 2.92.

Lesson 11:   The Decimal Expansion of Some Irrational Numbers

EUREKA
MATH

1. In which hundredth interval of the number line does $\sqrt{84}$ lie?

2. Determine the three-decimal digit approximation of the number $\sqrt{34}$.

3. Write the decimal expansion of $\sqrt{47}$ to at least two-decimal digits.

4. Write the decimal expansion of $\sqrt{46}$ to at least two-decimal digits.

5. Explain how to improve the accuracy of the decimal expansion of an irrational number.

6. Is the number $\sqrt{144}$ rational or irrational? Explain.

7. Is the number $0.\overline{64} = 0.646464646\ldots$ rational or irrational? Explain.

8. Henri computed the first 100 decimal digits of the number $\frac{352}{541}$ and got

    0.65064695009242144177449168207024029574861367837338262476894639556377079482439926062846580406654343807763401109057301294....

    He saw no repeating pattern to the decimal and so concluded that the number is irrational. Do you agree with Henri's conclusion? If not, what would you say to Henri?

9. Use a calculator to determine the decimal expansion of $\sqrt{35}$. Does the number appear to be rational or irrational? Explain.

10. Use a calculator to determine the decimal expansion of $\sqrt{101}$. Does the number appear to be rational or irrational? Explain.

11. Use a calculator to determine the decimal expansion of $\sqrt{7}$. Does the number appear to be rational or irrational? Explain.

12. Use a calculator to determine the decimal expansion of $\sqrt{8720}$. Does the number appear to be rational or irrational? Explain.

13. Use a calculator to determine the decimal expansion of $\sqrt{17956}$. Does the number appear to be rational or irrational? Explain.

14. Since the number $\frac{3}{5}$ is rational, must the number $\left(\frac{3}{5}\right)^2$ be rational as well? Explain.

15. If a number x is rational, must the number x^2 be rational as well? Explain.

16. Challenge: Determine the two-decimal digit approximation of the number $\sqrt[3]{9}$.

EUREKA MATH

Example 1

Find the decimal expansion of $\frac{35}{11}$.

Exercises 1–3

1. Find the decimal expansion of $\frac{5}{3}$ without using long division.

Lesson 12: Decimal Expansions of Fractions, Part 2

2. Find the decimal expansion of $\frac{5}{11}$ without using long division.

3. Find the decimal expansion of the number $\dfrac{23}{99}$ first without using long division and then again using long division.

Lesson Summary

For rational numbers, there is no need to guess and check in which interval of tenths, hundredths, or thousandths the number will lie.

For example, to determine where the fraction $\frac{1}{8}$ lies in the interval of tenths, compute using the following inequality:

$\frac{m}{10} < \frac{1}{8} < \frac{m+1}{10}$ Use the denominator of 10 because we need to find the tenths digit of $\frac{1}{8}$.

$m < \frac{10}{8} < m+1$ Multiply through by 10.

$m < 1\frac{1}{4} < m+1$ Simplify the fraction $\frac{10}{8}$.

The last inequality implies that $m = 1$ and $m + 1 = 2$ because $1 < 1\frac{1}{4} < 2$. Then, the tenths digit of the decimal expansion of $\frac{1}{8}$ is 1.

To find in which interval of hundredths $\frac{1}{8}$ lies, we seek consecutive integers m and $m + 1$ so that

$$\frac{1}{10} + \frac{m}{100} < \frac{1}{8} < \frac{1}{10} + \frac{m+1}{100}.$$

This is equivalent to

$$\frac{m}{100} < \frac{1}{8} - \frac{1}{10} < \frac{m+1}{100},$$

so we compute $\frac{1}{8} - \frac{1}{10} = \frac{2}{80} = \frac{1}{40}$. We have

$$\frac{m}{100} < \frac{1}{40} < \frac{m+1}{100}.$$

Multiplying through by 100 gives

$$m < \frac{10}{4} < m+1.$$

The last inequality implies that $m = 2$ and $m + 1 = 3$ because $2 < 2\frac{1}{2} < 3$. Then, the hundredths digit of the decimal expansion of $\frac{1}{8}$ is 2.

We can continue the process until the decimal expansion is complete or until we suspect a repeating pattern that we can verify.

Name _____ Date _____

Find the decimal expansion of $\frac{41}{6}$ without using long division.

1. Use rational approximation to determine the decimal expansion of $\frac{14}{9}$.

$$\frac{14}{9} = \frac{9}{9} + \frac{5}{9}$$

$$= 1 + \frac{5}{9}$$

> I know that $\frac{14}{9}$ lies between integers 1 and 2.

The ones digit is 1.

In the interval of tenths, we are looking for integers m and m + 1 so that

$$\frac{m}{10} < \frac{5}{9} < \frac{m+1}{10},$$

which is the same as

$$m < \frac{50}{9} < m + 1.$$

> Multiply every term by 10.

> Since I am looking for the tenths digit, I will use the denominator of 10 with my consecutive numerators of m and $m + 1$.

$$\frac{50}{9} = \frac{45}{9} + \frac{5}{9}$$

$$= 5 + \frac{5}{9}$$

The tenths digit is 5. The difference between $\frac{5}{9}$ and $\frac{5}{10}$ is

$$\frac{5}{9} - \frac{5}{10} = \frac{5}{90}.$$

In the interval of hundredths, we are looking for integers m and m + 1 so that

$$\frac{m}{100} < \frac{5}{90} < \frac{m+1}{100},$$

which is the same as

> I already did this when I was finding the tenths digit.

$$m < \frac{500}{90} < m + 1.$$

However, we already know that $\frac{500}{90} = \frac{50}{9} = 5 + \frac{5}{9}$; therefore, the hundredths digit is 5. Because we keep getting $\frac{5}{9}$, we can assume the digit of 5 will continue to repeat. Therefore, the decimal expansion of $\frac{14}{9}$ is 1.555….

2. Use rational approximation to determine the decimal expansion of $\frac{83}{37}$ to at least 3 decimal digits.

This problem is asking for the decimal expansion of $\frac{9}{37}$.

$$\frac{83}{37} = \frac{74}{37} + \frac{9}{37}$$
$$= 2 + \frac{9}{37}$$

I will use the same process as Problem 1 to find the tenths, hundredths, and thousandths digits.

The ones digit is 2.

In the interval of tenths, we are looking for integers m and $m+1$ so that

$$\frac{m}{10} < \frac{9}{37} < \frac{m+1}{10},$$

which is the same as

So far, I have 2.2 as my decimal expansion.

$$m < \frac{90}{37} < m+1.$$

$$\frac{90}{37} = \frac{74}{37} + \frac{16}{37} = 2 + \frac{16}{37}$$

The tenths digit is 2.

The difference between $\frac{9}{37}$ and $\frac{2}{10}$ is

I need to find the difference between $\frac{9}{37}$ and 0.2.

$$\frac{9}{37} - \frac{2}{10} = \frac{16}{370}.$$

In the interval of hundredths, we are looking for integers m and $m+1$ so that

$$\frac{m}{100} < \frac{16}{370} < \frac{m+1}{100},$$

which is the same as

So far, I have 2.24 as my decimal expansion.

$$m < \frac{1600}{370} < m+1.$$

$$\frac{1600}{370} = \frac{160}{37} = \frac{148}{37} + \frac{12}{37} = 4 + \frac{12}{37}$$

I need to find the difference between $\frac{9}{37}$ and the tenths and hundredths digits and use the result to find the thousandths digit.

The hundredths digit is 4.

The difference between $\frac{9}{37}$ and $\left(\frac{2}{10} + \frac{4}{100}\right)$ is

$$\frac{9}{37} - \left(\frac{2}{10} + \frac{4}{100}\right) = \frac{9}{37} - \frac{24}{100} = \frac{12}{3700}.$$

EUREKA
MATH

In the interval of thousandths, we are looking for integers m and $m + 1$ so that

$$\frac{m}{1000} < \frac{12}{3700} < \frac{m + 1}{1000},$$

which is the same as

$$m < \frac{12000}{3700} < m + 1.$$

$$\frac{12000}{3700} = \frac{120}{37} = \frac{111}{37} + \frac{9}{37} = 3 + \frac{9}{37}$$

> The fraction $\frac{9}{37}$ was where I started. The decimals will repeat from this point on.

The thousandths digit is 3.

We see again the fraction $\frac{9}{37}$, so we can expect the decimal digits to repeat at this point. Therefore, the decimal approximation of $\frac{83}{37}$ is $2.243243243....$

3. Use rational approximation to determine which number is larger, $\sqrt{12}$ or $\frac{11}{3}$.

The number $\sqrt{12}$ is between 3 and 4. In the sequence of tenths, $\sqrt{12}$ is between 3.4 and 3.5 because $3.4^2 < \left(\sqrt{12}\right)^2 < 3.5^2$. In the sequence of hundredths, $\sqrt{12}$ is between 3.46 and 3.47 because $3.46^2 < \left(\sqrt{12}\right)^2 < 3.47^2$. In the sequence of thousandths, $\sqrt{12}$ is between 3.464 and 3.465 because $3.464^2 < \left(\sqrt{12}\right)^2 < 3.465^2$. The decimal expansion of $\sqrt{12}$ is approximately $3.464....$

$$\frac{11}{3} = \frac{9}{3} + \frac{2}{3} = 3 + \frac{2}{3}$$

In the interval of tenths, we are looking for the integers m and $m + 1$ so that

$$\frac{m}{10} < \frac{2}{3} < \frac{m + 1}{10},$$

which is the same as

> I remember this type of problem from the last lesson.

$$m < \frac{20}{3} < m + 1.$$

$$\frac{20}{3} = \frac{18}{3} + \frac{2}{3} = 6 + \frac{2}{3}$$

The tenths digit is 6. Since the fraction $\frac{2}{3}$ has reappeared, we can assume that the next digit is also 6, and the work will continue to repeat. Therefore, the decimal expansion of $\frac{11}{3}$ is $3.666...$, and $\frac{11}{3} > \sqrt{12}$.

1. Without using long division, explain why the tenths digit of $\frac{3}{11}$ is a 2.

2. Find the decimal expansion of $\frac{25}{9}$ without using long division.

3. Find the decimal expansion of $\frac{11}{41}$ to at least 5 digits without using long division.

4. Which number is larger, $\sqrt{10}$ or $\frac{28}{9}$? Answer this question without using long division.

5. Sam says that $\frac{7}{11} = 0.63$, and Jaylen says that $\frac{7}{11} = 0.636$. Who is correct? Why?

Exploratory Challenge/Exercises 1–11

1. Rodney thinks that $\sqrt[3]{64}$ is greater than $\frac{17}{4}$. Sam thinks that $\frac{17}{4}$ is greater. Who is right and why?

2. Which number is smaller, $\sqrt[3]{27}$ or 2.89? Explain.

3. Which number is smaller, $\sqrt{121}$ or $\sqrt[3]{125}$? Explain.

4. Which number is smaller, $\sqrt{49}$ or $\sqrt[3]{216}$? Explain.

5. Which number is greater, $\sqrt{50}$ or $\dfrac{319}{45}$? Explain.

6. Which number is greater, $\dfrac{5}{11}$ or $0.\overline{4}$? Explain.

EUREKA
MATH

7. Which number is greater, $\sqrt{38}$ or $\dfrac{154}{25}$? Explain.

8. Which number is greater, $\sqrt{2}$ or $\dfrac{15}{9}$? Explain.

9. Place each of the following numbers at its approximate location on the number line: $\sqrt{25}, \sqrt{28}, \sqrt{30}, \sqrt{32}, \sqrt{35}$, and $\sqrt{36}$.

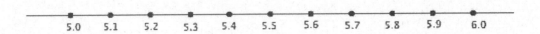

10. Challenge: Which number is larger, $\sqrt{5}$ or $\sqrt[3]{11}$?

11. A certain chessboard is being designed so that each square has an area of 3 in². What is the length of one edge of the board rounded to the tenths place? (A chessboard is composed of 64 squares as shown.)

Lesson Summary

Finding the first few places of the decimal expansion of numbers allows us to compare the numbers.

Name _____ Date _____

Place each of the following numbers at its approximate location on the number line: $\sqrt{12}$, $\sqrt{16}$, $\dfrac{20}{6}$, $3.\overline{53}$, and $\sqrt[3]{27}$.

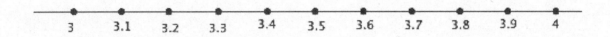

1. Which number is smaller, $\sqrt[3]{125}$ or $\sqrt{30}$? Explain your answer.

> I can use what I know about perfect squares to approximate $\sqrt{30}$.

$$\sqrt[3]{125} = \sqrt[3]{5^3} = 5$$

The number $\sqrt{30}$ is between 5 and 6 but definitely more than 5. Therefore, $\sqrt[3]{125} < \sqrt{30}$, and $\sqrt[3]{125}$ is smaller.

2. Which number is smaller, $\sqrt{64}$ or $\sqrt[3]{512}$? Explain your answer.

$$\sqrt{64} = \sqrt{8^2} = 8$$
$$\sqrt[3]{512} = \sqrt[3]{8^3} = 8$$

The numbers $\sqrt{64}$ and $\sqrt[2]{512}$ are equal because both are equal to 8.

3. Which number is larger, $\sqrt{68}$ or $\frac{829}{99}$? Explain your answer.

> I can use the method of rational approximation or long division to find the decimal expansion of $\frac{829}{99}$.

The number $\frac{829}{99}$ is equal to $8.\overline{37}$.

The number $\sqrt{68}$ is between 8 and 9 because $8^2 < \left(\sqrt{68}\right)^2 < 9^2$. The number $\sqrt{68}$ is between 8.2 and 8.3 because $8.2^2 < \left(\sqrt{68}\right)^2 < 8.3^2$. The approximate decimal value of $\sqrt{68}$ is 8.2.... Since $8.2 < 8.\overline{37}$, then $\sqrt{68} < \frac{829}{99}$. Thus, the fraction $\frac{829}{99}$ is larger.

> I only need to determine the decimal approximation for the tenths digit of $\sqrt{68}$ to determine which number is larger.

4. Which number is larger, $\frac{11}{15}$ or $0.\overline{732}$? Explain your answer.

The number $\frac{11}{15}$ is equal to $0.7\overline{3}$. Since $0.7\overline{3} > 0.\overline{732}...$, then $\frac{11}{15} > 0.\overline{732}$. Thus, the number $\frac{11}{15}$ is larger.

5. Which of the two right triangles shown below, measured in units, has the longer hypotenuse? Approximately how much longer is it?

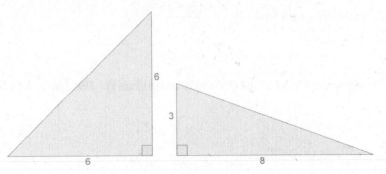

Let x represent the length of the hypotenuse of the triangle on the left.

$$6^2 + 6^2 = x^2$$
$$36 + 36 = x^2$$
$$72 = x^2$$
$$\sqrt{72} = \sqrt{x^2}$$
$$\sqrt{72} = x$$

> I need to use two different variables since I don't know the value of either or if they are the same number.

The number $\sqrt{72}$ is between 8 and 9 because $8^2 < \left(\sqrt{72}\right)^2 < 9^2$. The number $\sqrt{72}$ is between 8.4 and 8.5 because $8.4^2 < \left(\sqrt{72}\right)^2 < 8.5^2$. The number $\sqrt{72}$ is between 8.48 and 8.49 because $8.48^2 < \left(\sqrt{72}\right)^2 < 8.49^2$. The approximate decimal value of $\sqrt{72}$ is $8.48\ldots$.

Let y represent the length of the hypotenuse of the triangle on the right.

$$3^2 + 8^2 = y^2$$
$$9 + 64 = y^2$$
$$73 = y^2$$
$$\sqrt{73} = \sqrt{y^2}$$
$$\sqrt{73} = y$$

> I know that $\sqrt{72} < \sqrt{73}$, but the question is asking approximately how much longer is $\sqrt{73}$ than $\sqrt{72}$, so I need to determine the decimal approximation for y.

The number $\sqrt{73}$ is between 8 and 9 because $8^2 < \left(\sqrt{73}\right)^2 < 9^2$. The number $\sqrt{73}$ is between 8.5 and 8.6 because $8.5^2 < \left(\sqrt{73}\right)^2 < 8.6^2$. The number $\sqrt{73}$ is between 8.54 and 8.55 because $8.54^2 < \left(\sqrt{73}\right)^2 < 8.55^2$. The approximate decimal value of $\sqrt{73}$ is $8.54\ldots$.

The triangle on the right has the longer hypotenuse. It is approximately 0.06 units longer than the hypotenuse of the triangle on the left.

EUREKA MATH

1. Which number is smaller, $\sqrt[3]{343}$ or $\sqrt{48}$? Explain.

2. Which number is smaller, $\sqrt{100}$ or $\sqrt[3]{1000}$? Explain.

3. Which number is larger, $\sqrt{87}$ or $\dfrac{929}{99}$? Explain.

4. Which number is larger, $\dfrac{9}{13}$ or $0.\overline{692}$? Explain.

5. Which number is larger, 9.1 or $\sqrt{82}$? Explain.

6. Place each of the following numbers at its approximate location on the number line: $\sqrt{144}$, $\sqrt[3]{1000}$, $\sqrt{130}$, $\sqrt{110}$, $\sqrt{120}$, $\sqrt{115}$, and $\sqrt{133}$. Explain how you knew where to place the numbers.

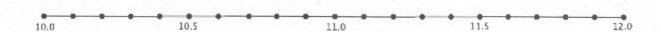

7. Which of the two right triangles shown below, measured in units, has the longer hypotenuse? Approximately how much longer is it?

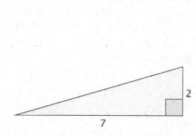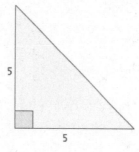

Opening Exercise

a.   Write an equation for the area, A, of the circle shown.

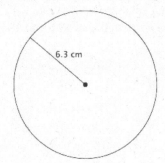

6.3 cm

b.   Write an equation for the circumference, C, of the circle shown.

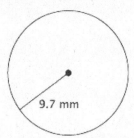

9.7 mm

c.   Each of the squares in the grid on the following page has an area of 1 unit².

   i.   Estimate the area of the circle shown by counting squares.

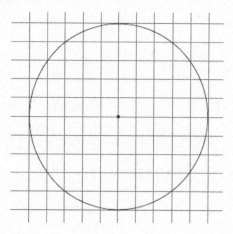

   ii.   Calculate the area of the circle using a radius of 5 units.
         Use 3.14 as an approximation for π.

Exercises 1–4

1. Gerald and Sarah are building a wheel with a radius of 6.5 cm and are trying to determine the circumference. Gerald says, "Because $6.5 \times 2 \times 3.14 = 40.82$, the circumference is 40.82 cm." Sarah says, "Because $6.5 \times 2 \times 3.10 = 40.3$ and $6.5 \times 2 \times 3.21 = 41.73$, the circumference is somewhere between 40.3 and 41.73." Explain the thinking of each student.

2. Estimate the value of the number $(6.12486...)^2$.

3. Estimate the value of the number $(9.204107...)^2$.

4. Estimate the value of the number $(4.014325...)^2$.

EUREKA
MATH

Lesson Summary

Numbers, such as π, are frequently approximated in order to compute with them. Common approximations for π are 3.14 and $\frac{22}{7}$. It should be understood that using an approximate value of a number for computations produces an answer that is accurate to only the first few decimal digits.

10 by 10 Grid

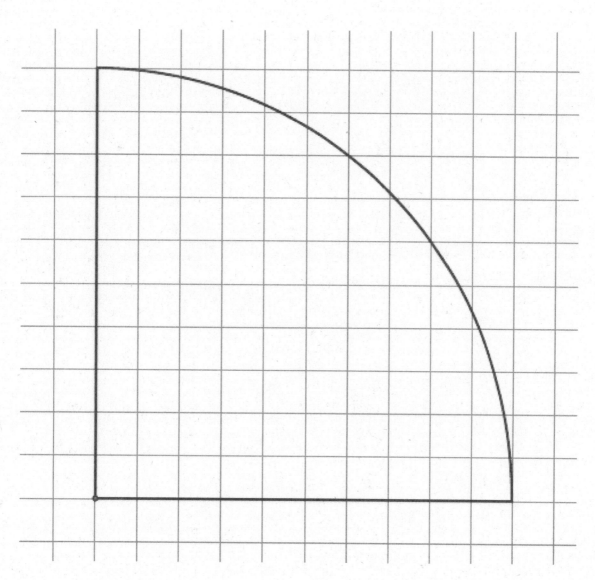

20 by 20 Grid

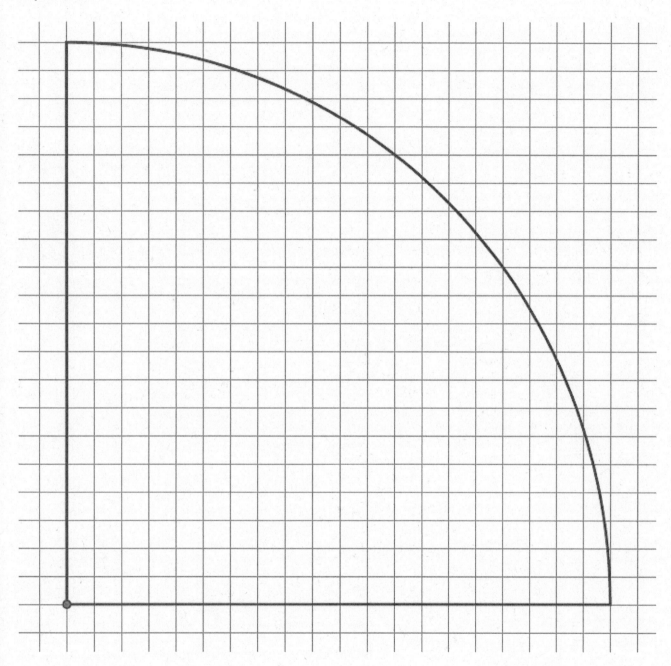

EUREKA
MATH®

© 2019 Great Minds®. eureka-math.org

Name _____   Date _____

Describe how we found a decimal approximation for π.

1. Carrie estimated π to be $3.10 < \pi < 3.21$. If she uses this approximation of π to determine the area of a circle with a radius of 7 in., what could the area be?

   The area of the circle with radius 7 in. will be between 151.9 in² and 157.29 in².

   > I need to substitute the values for π and the radius into the formula for the area of a circle, namely $A = \pi r^2$.

2. Winston estimated the circumference of a circle with a radius of 8.3 in. to be 52.29 in. What approximate value of π did he use? Is it an acceptable approximation of π? Explain.

$$C = 2\pi r$$
$$52.29 = 2\pi(8.3)$$
$$52.29 = 16.6\pi$$
$$\frac{52.29}{16.6} = \pi$$
$$3.15 = \pi$$

   > I need to solve for π using properties of equality.

   Winston used 3.15 to approximate π. This is an acceptable approximation for π because it is in the interval that we approximated π to be in the lesson, $3.10 < \pi < 3.21$.

3. Estimate the value of the irrational number $(2.3856...)^2$.

   > I need to add 0.0001 to 2.3856 and then square it.

$$2.3856^2 < (2.3856...)^2 < 2.3857^2$$
$$5.69108736 < (2.3856...)^2 < 5.69156449$$

   $(2.3856...)^2 = 5.691$ **is correct up to three decimal digits.**

4. Estimate the value of the irrational number $(0.956321...)^2$.

   > I need to count how many decimal places the values are exactly the same.

$$0.956321^2 < (0.956321...)^2 < 0.956322^2$$
$$0.914549855 < (0.956321...)^2 < 0.9145517977$$

   $(0.956321...)^2 = 0.9145$ **is correct up to four decimal digits.**

1. Caitlin estimated π to be $3.10 < \pi < 3.21$. If she uses this approximation of π to determine the area of a circle with a radius of 5 cm, what could the area be?

2. Myka estimated the circumference of a circle with a radius of 4.5 in. to be 28.44 in. What approximate value of π did she use? Is it an acceptable approximation of π? Explain.

3. A length of ribbon is being cut to decorate a cylindrical cookie jar. The ribbon must be cut to a length that stretches the length of the circumference of the jar. There is only enough ribbon to make one cut. When approximating π to calculate the circumference of the jar, which number in the interval $3.10 < \pi < 3.21$ should be used? Explain.

4. Estimate the value of the number $(1.86211...)^2$.

5. Estimate the value of the number $(5.9035687...)^2$.

6. Estimate the value of the number $(12.30791...)^2$.

7. Estimate the value of the number $(0.6289731...)^2$.

8. Estimate the value of the number $(1.112223333...)^2$.

9. Which number is a better estimate for π, $\frac{22}{7}$ or 3.14? Explain.

10. To how many decimal digits can you correctly estimate the value of the number $(4.56789012...)^2$?

Proof of the Pythagorean Theorem

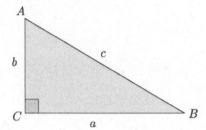

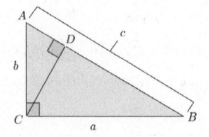

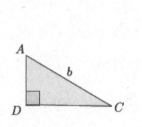 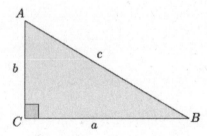 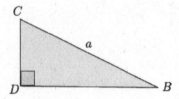

EUREKA MATH®

Discussion

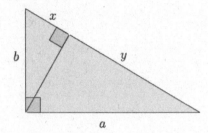

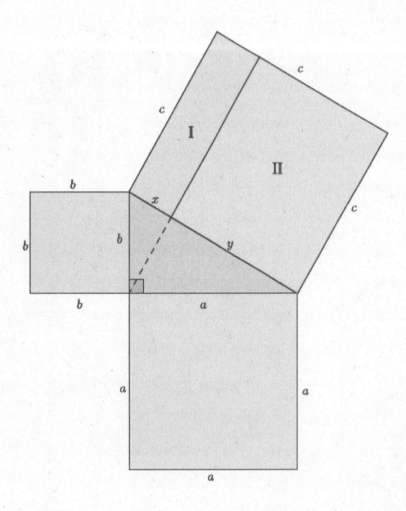

EUREKA
MATH®

Lesson Summary

The Pythagorean theorem can be proven by drawing an altitude in the given right triangle and identifying three similar triangles. We can see geometrically how the large square drawn on the hypotenuse of the triangle has an area summing to the areas of the two smaller squares drawn on the legs of the right triangle.

Name _____   Date _____

Explain a proof of the Pythagorean theorem in your own words. Use diagrams and concrete examples, as necessary, to support your explanation.

1. For the right triangle shown below, identify and use similar triangles to illustrate the Pythagorean theorem.

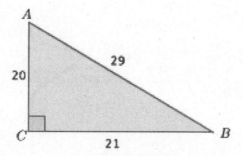

First, draw a segment that is perpendicular to side $\overline{AB}$ that goes through point C. The point of intersection of that segment and side $\overline{AB}$ will be marked as point D.

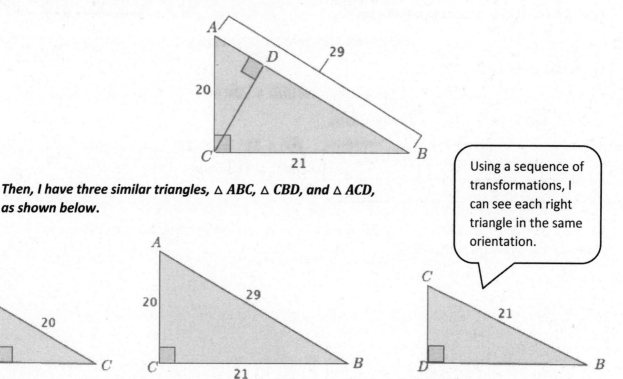

Then, I have three similar triangles, △ ABC, △ CBD, and △ ACD, as shown below.

Using a sequence of transformations, I can see each right triangle in the same orientation.

△ ABC and △ CBD are similar because each one has a right angle, and they both share ∠B. By AA criterion, △ ABC ~ △ CBD.

△ ABC and △ ACD are similar because each one has a right angle, and they both share ∠A. By AA criterion, △ ABC ~ △ ACD.

By the transitive property, we also know that $\triangle ACD \sim \triangle CBD$.

Since the triangles are similar, they have corresponding sides that are equal in ratio.

For $\triangle ABC$ and $\triangle CBD$,

$$\frac{21}{29} = \frac{|BD|}{21},$$

which is the same as $21^2 = 29(|BD|)$.

For $\triangle ABC$ and $\triangle ACD$,

$$\frac{20}{29} = \frac{|AD|}{20},$$

which is the same as $20^2 = 29(|AD|)$.

> Because my right triangles are in the same orientation, I can make the following proportion using $\triangle ABC$ and $\triangle CBD$:
> $$\frac{\text{base}}{\text{hypotenuse}} = \frac{\text{base}}{\text{hypotenuse}}.$$
> I can solve any proportion by using cross multiplication.

Adding these two equations together, I get

$$21^2 + 20^2 = 29(|BD|) + 29(|AD|).$$

By the distributive property,

$$21^2 + 20^2 = 29(|BD| + |AD|);$$

however, $|BD| + |AD| = |AB| = 29$. Therefore,

> I found this by looking at my original triangle and the triangle where the altitude was drawn.

$$20^2 + 21^2 = 29(29)$$
$$20^2 + 21^2 = 29^2.$$

2. For the right triangle shown below, identify and use squares formed by the sides of the triangle to illustrate the Pythagorean theorem.

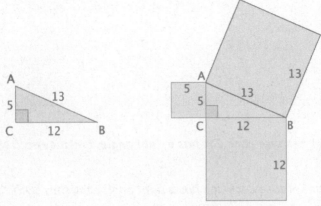

The sum of the areas of the smallest squares is $5^2 + 12^2$, which is 169. The area of the largest square is 13^2, which is 169. The sum of the areas of the squares of the legs is equal to the area of the square of the hypotenuse; therefore, for legs a and b, and hypotenuse c, we see that $a^2 + b^2 = c^2$.

Lesson 15:   Pythagorean Theorem, Revisited

EUREKA MATH

3. Can any similar figures be drawn off the sides of the right triangle to prove the Pythagorean theorem? Use computations to show that the sum of the areas of the figures off of the sides a and b equals the area of the figure off of side c.

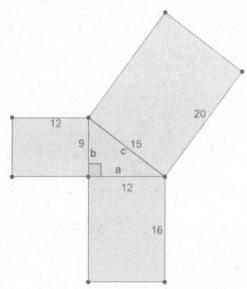

The rectangles are similar because their corresponding side lengths are equal in scale factor.

$$\frac{9}{12} = \frac{12}{16} = \frac{15}{20} = 0.75$$

The area of the smaller rectangles are 108 square units and 192 square units, and the area of the larger rectangle is 300 square units. The sum of the smaller areas is equal to the larger area.

$$108 + 192 = 300$$
$$300 = 300$$

Therefore, the sum of the areas of the smaller similar rectangles does equal the area of the larger similar rectangle proving the Pythagorean theorem with similar figures.

4. The following image for the Pythagorean theorem contains an error. Explain what is wrong.

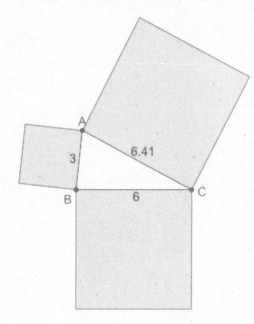

Based on the proof shown in class, we would expect the sum of the two smaller areas to be equal to the larger area. The smaller areas are 9 and 36, while the larger area is 41.0881. That is, $9 + 36$ should equal 41.0881. However, $9 + 36 = 45$.

We know that the Pythagorean theorem only works for right triangles. Considering the converse of the Pythagorean theorem, when we use the given side lengths, we do not get a true statement.

$$3^2 + 6^2 = 6.41^2$$
$$9 + 36 = 41.0881$$
$$45 \neq 41.0881$$

Therefore, the original triangle is not a right triangle, so it makes sense that the areas of the squares were not adding up like we expected.

1. For the right triangle shown below, identify and use similar triangles to illustrate the Pythagorean theorem.

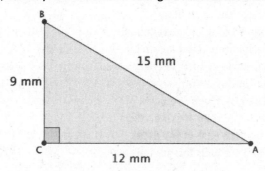

2. For the right triangle shown below, identify and use squares formed by the sides of the triangle to illustrate the Pythagorean theorem.

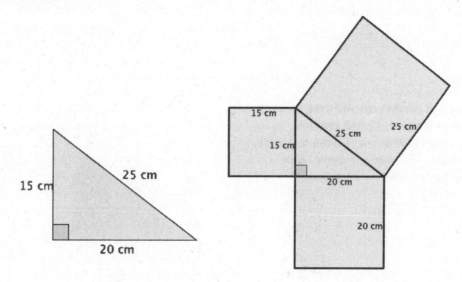

3. Reese claimed that any figure can be drawn off the sides of a right triangle and that as long as they are similar figures, then the sum of the areas off of the legs will equal the area off of the hypotenuse. She drew the diagram by constructing rectangles off of each side of a known right triangle. Is Reese's claim correct for this example? In order to prove or disprove Reese's claim, you must first show that the rectangles are similar. If they are, then you can use computations to show that the sum of the areas of the figures off of the sides a and b equals the area of the figure off of side c.

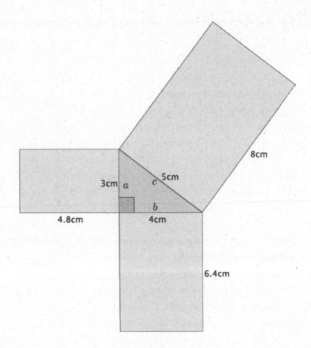

4. After learning the proof of the Pythagorean theorem using areas of squares, Joseph got really excited and tried explaining it to his younger brother. He realized during his explanation that he had done something wrong. Help Joseph find his error. Explain what he did wrong.

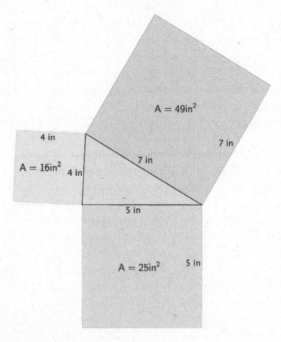

5. Draw a right triangle with squares constructed off of each side that Joseph can use the next time he wants to show his younger brother the proof of the Pythagorean theorem.

6. Explain the meaning of the Pythagorean theorem in your own words.

7. Draw a diagram that shows an example illustrating the Pythagorean theorem.

EUREKA
MATH

Proof of the Converse of the Pythagorean Theorem

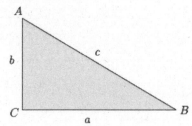
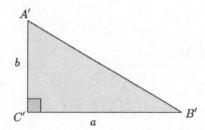

Exercises 1–7

1. Is the triangle with leg lengths of 3 mi. and 8 mi. and hypotenuse of length $\sqrt{73}$ mi. a right triangle? Show your work, and answer in a complete sentence.

2. What is the length of the unknown side of the right triangle shown below? Show your work, and answer in a complete sentence. Provide an exact answer and an approximate answer rounded to the tenths place.

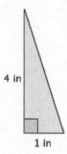

4 in

1 in

3. What is the length of the unknown side of the right triangle shown below? Show your work, and answer in a complete sentence. Provide an exact answer and an approximate answer rounded to the tenths place.

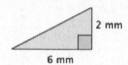

2 mm

6 mm

4. Is the triangle with leg lengths of 9 in. and 9 in. and hypotenuse of length $\sqrt{175}$ in. a right triangle? Show your work, and answer in a complete sentence.

EUREKA
MATH

5. Is the triangle with leg lengths of $\sqrt{28}$ cm and 6 cm and hypotenuse of length 8 cm a right triangle? Show your work, and answer in a complete sentence.

6. What is the length of the unknown side of the right triangle shown below? Show your work, and answer in a complete sentence.

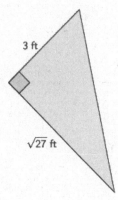

3 ft

$\sqrt{27}$ ft

7. The triangle shown below is an isosceles right triangle. Determine the length of the legs of the triangle. Show your work, and answer in a complete sentence.

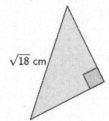

$\sqrt{18}$ cm

Lesson Summary

The converse of the Pythagorean theorem states that if a triangle with side lengths a, b, and c satisfies $a^2 + b^2 = c^2$, then the triangle is a right triangle.

The converse can be proven using concepts related to congruence.

Name _____   Date _____

1. Is the triangle with leg lengths of 7 mm and 7 mm and a hypotenuse of length 10 mm a right triangle? Show your work, and answer in a complete sentence.

2. What would the length of the hypotenuse need to be so that the triangle in Problem 1 would be a right triangle? Show work that leads to your answer.

3. If one of the leg lengths is 7 mm, what would the other leg length need to be so that the triangle in Problem 1 would be a right triangle? Show work that leads to your answer.

1. What is the length of the unknown side of the right triangle shown below? Show your work, and answer in a complete sentence. Provide an exact answer and an approximate answer rounded to the tenths place.

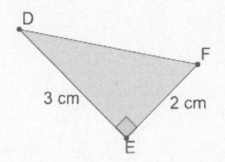

Let c cm represent the hypotenuse of the triangle.

$$3^2 + 2^2 = c^2$$
$$9 + 4 = c^2$$
$$13 = c^2$$
$$\sqrt{13} = \sqrt{c^2}$$
$$3.6 \approx c$$

> To estimate, I need the two perfect squares that surround 13, which are 9 and 16. 13 is slightly closer to 16 than 9, so $\sqrt{13}$ is closer to $\sqrt{16} = 4$ than $\sqrt{9} = 3$.

The hypotenuse of the triangle is exactly $\sqrt{13}$ cm and approximately 3.6 cm.

2. Is the triangle with leg lengths of $\sqrt{5}$ cm and 7 cm and hypotenuse of length $\sqrt{54}$ cm a right triangle? Show your work, and answer in a complete sentence.

> To simplify a square root that is squared, I need to remember the following:
> $$\left(\sqrt{5}\right)^2 = \sqrt{5} \cdot \sqrt{5}$$
> $$\left(\sqrt{5}\right)^2 = \sqrt{25}$$
> $$\left(\sqrt{5}\right)^2 = 5$$

$$\left(\sqrt{5}\right)^2 + 7^2 = \left(\sqrt{54}\right)^2$$
$$5 + 49 = 54$$
$$54 = 54$$

Yes, the triangle with leg lengths of $\sqrt{5}$ cm and 7 cm and hypotenuse of length $\sqrt{54}$ cm is a right triangle because the lengths satisfy the Pythagorean theorem.

3. Is the triangle with leg lengths of $\sqrt{8}$ cm and 10 cm and hypotenuse of length $\sqrt{164}$ cm a right triangle? Show your work, and answer in a complete sentence.

$$\left(\sqrt{8}\right)^2 + 10^2 = \left(\sqrt{164}\right)^2$$
$$8 + 100 = 164$$
$$108 \neq 164$$

No, the triangle with leg lengths of $\sqrt{8}$ cm and 10 cm and hypotenuse of length $\sqrt{164}$ cm is not a right triangle because the lengths do not satisfy the Pythagorean theorem.

1. What is the length of the unknown side of the right triangle shown below? Show your work, and answer in a complete sentence. Provide an exact answer and an approximate answer rounded to the tenths place.

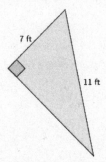

2. What is the length of the unknown side of the right triangle shown below? Show your work, and answer in a complete sentence. Provide an exact answer and an approximate answer rounded to the tenths place.

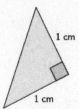

3. Is the triangle with leg lengths of $\sqrt{3}$ cm and 9 cm and hypotenuse of length $\sqrt{84}$ cm a right triangle? Show your work, and answer in a complete sentence.

4. Is the triangle with leg lengths of $\sqrt{7}$ km and 5 km and hypotenuse of length $\sqrt{48}$ km a right triangle? Show your work, and answer in a complete sentence.

5. What is the length of the unknown side of the right triangle shown below? Show your work, and answer in a complete sentence. Provide an exact answer and an approximate answer rounded to the tenths place.

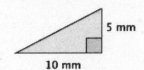

6. Is the triangle with leg lengths of 3 and 6 and hypotenuse of length $\sqrt{45}$ a right triangle? Show your work, and answer in a complete sentence.

7. What is the length of the unknown side of the right triangle shown below? Show your work, and answer in a complete sentence. Provide an exact answer and an approximate answer rounded to the tenths place.

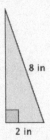

8 in

2 in

8. Is the triangle with leg lengths of 1 and $\sqrt{3}$ and hypotenuse of length 2 a right triangle? Show your work, and answer in a complete sentence.

9. Corey found the hypotenuse of a right triangle with leg lengths of 2 and 3 to be $\sqrt{13}$. Corey claims that since $\sqrt{13} = 3.61$ when estimating to two decimal digits, that a triangle with leg lengths of 2 and 3 and a hypotenuse of 3.61 is a right triangle. Is he correct? Explain.

10. Explain a proof of the Pythagorean theorem.

11. Explain a proof of the converse of the Pythagorean theorem.

Lesson 16:   Converse of the Pythagorean Theorem

EUREKA
MATH

Example 1

What is the distance between the two points A and B on the coordinate plane?

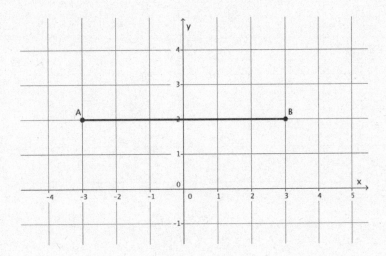

What is the distance between the two points A and B on the coordinate plane?

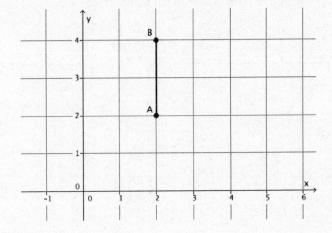

What is the distance between the two points A and B on the coordinate plane? Round your answer to the tenths place.

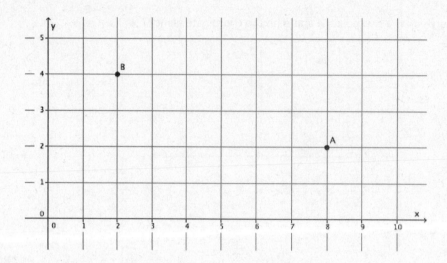

Example 2

Given two points A and B on the coordinate plane, determine the distance between them. First, make an estimate; then, try to find a more precise answer. Round your answer to the tenths place.

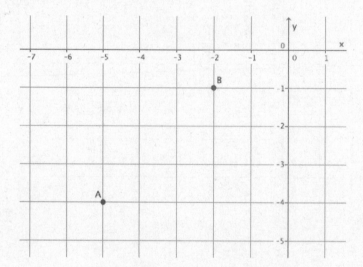

Lesson 17: Distance on the Coordinate Plane

EUREKA MATH

Exercises 1–4

For each of the Exercises 1–4, determine the distance between points A and B on the coordinate plane. Round your answer to the tenths place.

1.

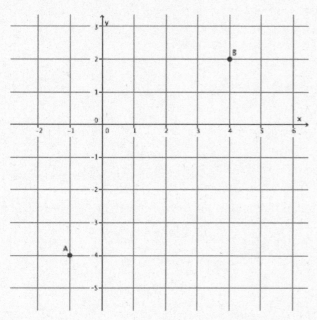

2.

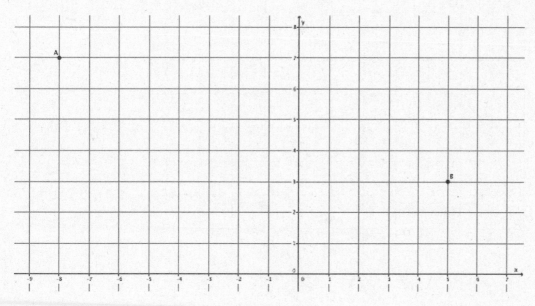

3.

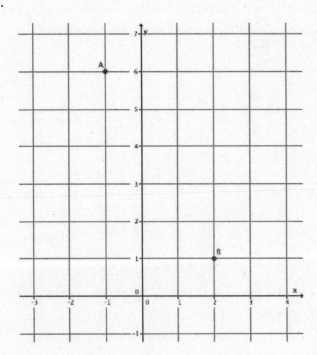

4.

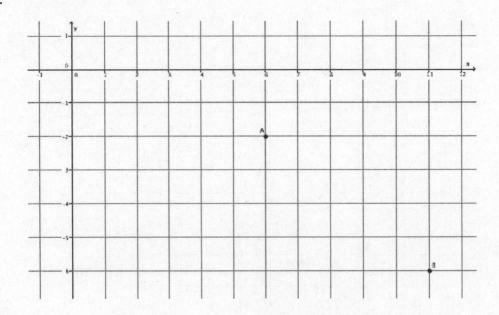

Lesson 17:   Distance on the Coordinate Plane

© 2019 Great Minds®. eureka-math.org

EUREKA
MATH

Example 3

Is the triangle formed by the points A, B, C a right triangle?

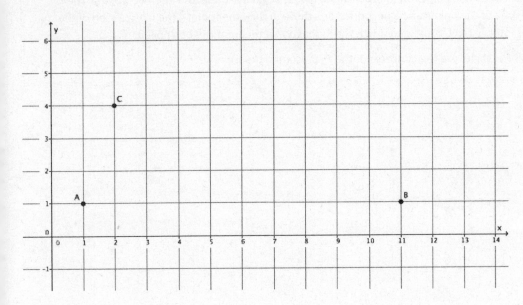

Lesson Summary

To determine the distance between two points on the coordinate plane, begin by connecting the two points. Then, draw a vertical line through one of the points and a horizontal line through the other point. The intersection of the vertical and horizontal lines forms a right triangle to which the Pythagorean theorem can be applied.

To verify if a triangle is a right triangle, use the converse of the Pythagorean theorem.

Lesson 17:   Distance on the Coordinate Plane

Name _____   Date _____

Use the following diagram to answer the questions below.

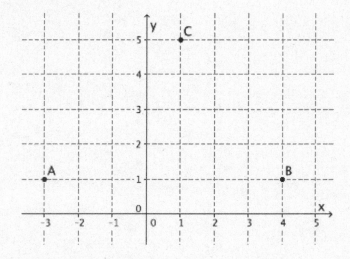

1. Determine $|AC|$. Leave your answer in square root form unless it is a perfect square.

2. Determine $|CB|$. Leave your answer in square root form unless it is a perfect square.

3. Is the triangle formed by the points A, B, C a right triangle? Explain why or why not.

1. Determine the distance between points A and B on the coordinate plane. Round your answer to the tenths place.

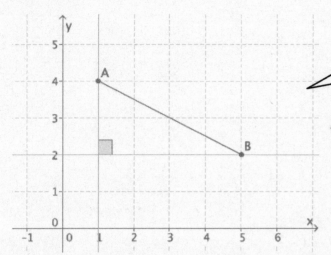

To determine the distance between the points, I am going to use the grid lines from the coordinate plane to create a right triangle.

Let c represent |AB|.

$$2^2 + 4^2 = c^2$$
$$4 + 16 = c^2$$
$$20 = c^2$$
$$\sqrt{20} = c$$
$$4.5 \approx c$$

The distance between points A and B is about 4.5 units.

2. Is the triangle formed by points A, B, and C a right triangle?

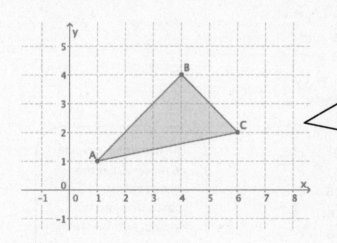

None of these sides are horizontal or vertical, so I cannot simply count to find their lengths.

I need to create right triangles to find the lengths of the sides and then use those lengths in the Pythagorean theorem.

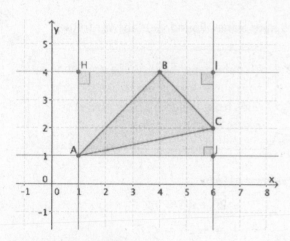

Let c represent $|AB|$.

$$3^2 + 3^2 = c^2$$
$$9 + 9 = c^2$$
$$18 = c^2$$
$$\sqrt{18} = \sqrt{c^2}$$
$$\sqrt{18} = c$$

Let c represent $|BC|$.

$$2^2 + 2^2 = c^2$$
$$4 + 4 = c^2$$
$$8 = c^2$$
$$\sqrt{8} = \sqrt{c^2}$$
$$\sqrt{8} = c$$

Let c represent $|AC|$.

$$5^2 + 1^2 = c^2$$
$$25 + 1 = c^2$$
$$26 = c^2$$
$$\sqrt{26} = \sqrt{c^2}$$
$$\sqrt{26} = c$$

The hypotenuse is side $\overline{AC}$ since $\sqrt{26}$ is the longest side. I substituted into the Pythagorean theorem and got a true statement, meaning I have a right triangle.

$$\left(\sqrt{18}\right)^2 + \left(\sqrt{8}\right)^2 = \left(\sqrt{26}\right)^2$$
$$18 + 8 = 26$$
$$26 = 26$$

Yes, the points do form a right triangle.

EUREKA MATH

For each of the Problems 1–4, determine the distance between points A and B on the coordinate plane. Round your answer to the tenths place.

1.

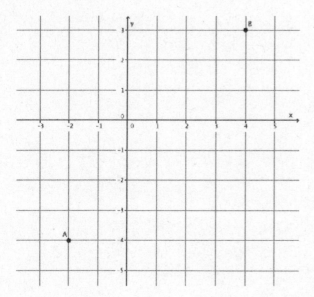

2.

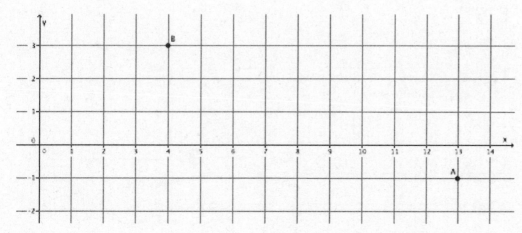

EUREKA
MATH

© 2019 Great Minds®. eureka-math.org

3.

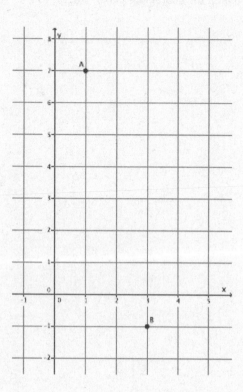

4.

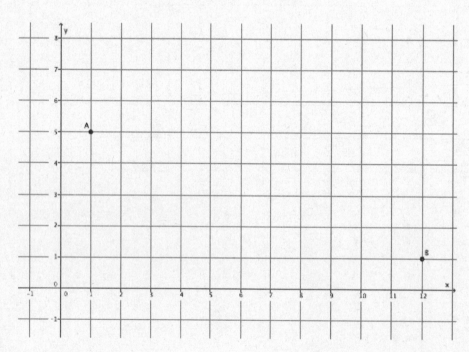

Lesson 17:   Distance on the Coordinate Plane

EUREKA
MATH

5. Is the triangle formed by points A, B, C a right triangle?

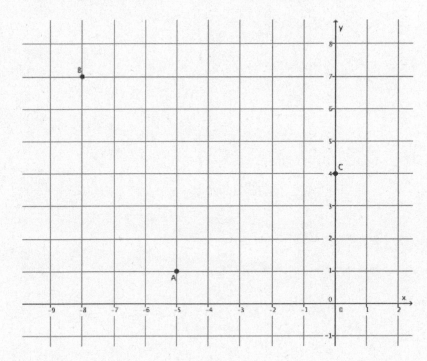

Exercises

1. The area of the right triangle shown below is 26.46 in^2. What is the perimeter of the right triangle? Round your answer to the tenths place.

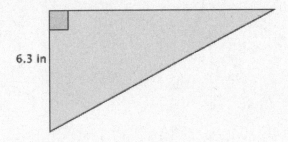

6.3 in

© 2019 Great Minds®. eureka-math.org

2. The diagram below is a representation of a soccer goal.

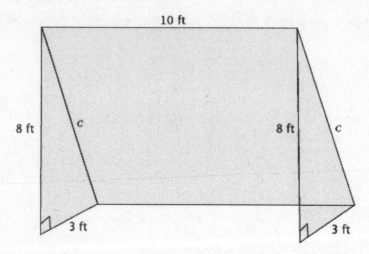

a. Determine the length of the bar, c, that would be needed to provide structure to the goal. Round your answer to the tenths place.

b. How much netting (in square feet) is needed to cover the entire goal?

3. The typical ratio of length to width that is used to produce televisions is 4: 3.

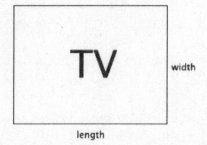

width

length

A TV with length 20 inches and width 15 inches, for example, has sides in a 4: 3 ratio; as does any TV with length $4x$ inches and width $3x$ inches for any number x.

a. What is the advertised size of a TV with length 20 inches and width 15 inches?

b. A 42" TV was just given to your family. What are the length and width measurements of the TV?

c.   Check that the dimensions you got in part (b) are correct using the Pythagorean theorem.

d.   The table that your TV currently rests on is 30" in length. Will the new TV fit on the table? Explain.

4.   Determine the distance between the following pairs of points. Round your answer to the tenths place. Use graph paper if necessary.

a. $(7, 4)$ and $(-3, -2)$

b. $(-5, 2)$ and $(3, 6)$

EUREKA
MATH

c.   Challenge: (x_1, y_1) and (x_2, y_2). Explain your answer.

5. What length of ladder is needed to reach a height of 7 feet along the wall when the base of the ladder is 4 feet from the wall? Round your answer to the tenths place.

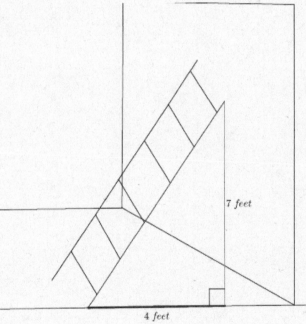

7 feet

4 feet

Name _____ Date _____

Use the diagram of the equilateral triangle shown below to answer the following questions. Show the work that leads to your answers.

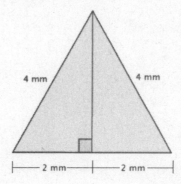

a. What is the perimeter of the triangle?

b. What is the height, h mm, of the equilateral triangle? Write an exact answer using a square root and an approximate answer rounded to the tenths place.

c. Using the approximate height found in part (b), estimate the area of the equilateral triangle.

1. A 55 in. TV is advertised on sale at a local store. What are the length and width of the television?

    Let x be the factor applied to the ratio 4: 3.

    $$(4x)^2 + (3x)^2 = 55^2$$
    $$16x^2 + 9x^2 = 3025$$
    $$25x^2 = 3025$$
    $$\frac{25}{25}x^2 = \frac{3025}{25}$$
    $$x^2 = 121$$
    $$\sqrt{x^2} = \sqrt{121}$$
    $$x = 11$$

    > I remember from the notes that the dimensions of a television are in the ratio 4: 3 and that the size of the television is actually the length of the diagonal. Therefore, $4x$ and $3x$ are the legs of the right triangle while 55 is the hypotenuse.

    Since x = 11, 3x = 33 and 4x = 44. Therefore, the dimensions of the TV are 44 *in. by* 33 *in.*

2. The soccer team was instructed to run the perimeter of their soccer field, which has dimensions of 115 yards by 74 yards. One player decided to run the length and width and then finished by running diagonally. To the nearest tenth of a yard, how many fewer yards of running did this player complete than the rest of the team?

    $$P = 2l + 2w$$
    $$P = 2(115) + 2(74)$$
    $$P = 230 + 148$$
    $$P = 378$$

    The team ran 378 *yards.*

    Let a yards represent the length of the field, b yards represent the width of the field, and c yards represent the diagonal length of the field.

    $$a^2 + b^2 = c^2$$
    $$115^2 + 74^2 = c^2$$
    $$13225 + 5476 = c^2$$
    $$18701 = c^2$$
    $$\sqrt{18701} = \sqrt{c^2}$$
    $$136.8 \approx c$$

    > I need to find the diagonal length of the field and then use it to find the total distance the single player ran. The number $\sqrt{18,701}$ is between 136 and 137. In the sequence of tenths, the number is between 136.7 and 136.8 because $136.7^2 < \left(\sqrt{18,701}\right)^2 < 136.8^2$. Since 18,701 is closer to 136.8^2 than 136.7^2, the approximate length of the hypotenuse is 136.8 yards.

$$115 + 74 + 136.8 = 325.8$$

The player ran approximately 325.8 yards.

$$378 - 325.8 = 52.2$$

The player ran approximately 52.2 fewer yards than the rest of the team.

3. The area of the right triangle shown below is 51.24 in^2.

   a. What is the height of the triangle?

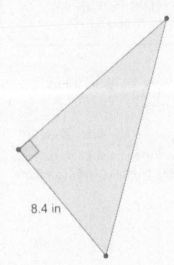

8.4 in

Let h in. *represent the height of the triangle.*

$$A = \frac{1}{2}bh$$

$$51.24 = \frac{1}{2}(8.4)h$$

$$51.24 = 4.2h$$

$$\frac{51.24}{4.2} = \frac{4.2}{4.2}h$$

$$12.2 = h$$

The height of the triangle is 12.2 in.

   b. What is the perimeter of the right triangle? Round your answer to the tenths place.

   Let c in. *represent the length of the hypotenuse.*

$$8.4^2 + 12.2^2 = c^2$$

$$70.56 + 148.84 = c^2$$

$$219.4 = c^2$$

$$\sqrt{219.4} = \sqrt{c^2}$$

$$14.8 \approx c$$

> In order to find the perimeter, I first need all three sides. To find the hypotenuse, I can use the Pythagorean theorem.

The number $\sqrt{219.4}$ is between 14 and 15. In the sequence of tenths, the number is between 14.8 and 14.9 because $14.8^2 < \left(\sqrt{219.4}\right)^2 < 14.9^2$. Since 219.4 is closer to 14.8^2 than 14.9^2, the approximate length of the hypotenuse is 14.8 in.

$$8.4 + 12.2 + 14.8 = 35.4$$

The perimeter of the triangle is approximately 35.4 in.

1. A 70" TV is advertised on sale at a local store. What are the length and width of the television?

2. There are two paths that one can use to go from Sarah's house to James' house. One way is to take C Street, and the other way requires you to use A Street and B Street. How much shorter is the direct path along C Street?

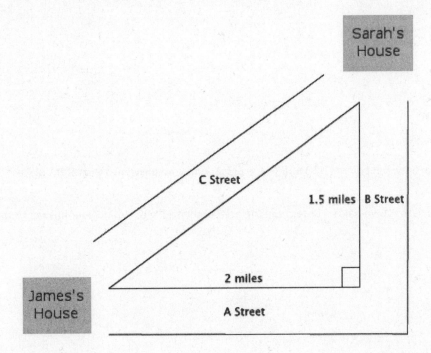

3. An isosceles right triangle refers to a right triangle with equal leg lengths, *s*, as shown below.

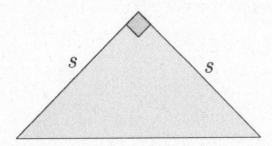

What is the length of the hypotenuse of an isosceles right triangle with a leg length of 9 cm? Write an exact answer using a square root and an approximate answer rounded to the tenths place.

4. The area of the right triangle shown to the right is 66.5 cm².

    a. What is the height of the triangle?

    b. What is the perimeter of the right triangle? Round your answer to the tenths place.

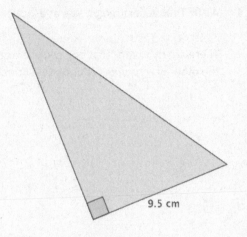

9.5 cm

5. What is the distance between points $(1, 9)$ and $(-4, -1)$? Round your answer to the tenths place.

6. An equilateral triangle is shown below. Determine the area of the triangle. Round your answer to the tenths place.

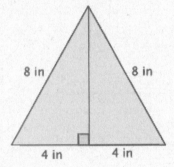

8 in   8 in

4 in   4 in

EUREKA
MATH®

Exercises 1–2

Note: Figures not drawn to scale.

1. Determine the volume for each figure below.

   a. Write an expression that shows volume in terms of the area of the base, B, and the height of the figure. Explain the meaning of the expression, and then use it to determine the volume of the figure.

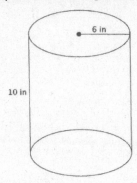

   b. Write an expression that shows volume in terms of the area of the base, B, and the height of the figure. Explain the meaning of the expression, and then use it to determine the volume of the figure.

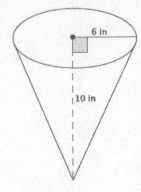

2.

  a. Write an expression that shows volume in terms of the area of the base, B, and the height of the figure.
      Explain the meaning of the expression, and then use it to determine the volume of the figure.

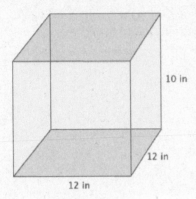

  b. The volume of the square pyramid shown below is 480 in^3. What might be a reasonable guess for the formula
      for the volume of a pyramid? What makes you suggest your particular guess?

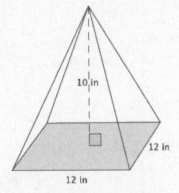

EUREKA
MATH

Example 1

State as many facts as you can about a cone.

Exercises 3–10

3. What is the lateral length (slant height) of the cone shown below?

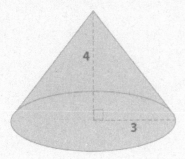

4. Determine the exact volume of the cone shown below.

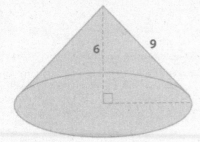

5. What is the lateral length (slant height) of the pyramid shown below? Give an exact square root answer and an approximate answer rounded to the tenths place.

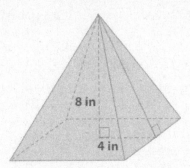

8 in

4 in

6. Determine the volume of the square pyramid shown below. Give an exact answer using a square root.

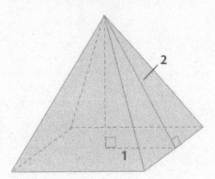

2

1

7. What is the length of the chord of the sphere shown below? Give an exact answer using a square root.

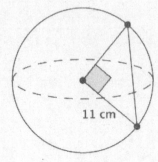

11 cm

EUREKA
MATH®

8. What is the length of the chord of the sphere shown below? Give an exact answer using a square root.

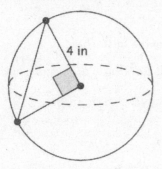

9. What is the volume of the sphere shown below? Give an exact answer using a square root.

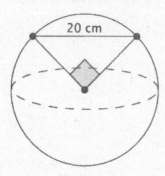

10. What is the volume of the sphere shown below? Give an exact answer using a square root.

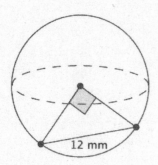

Lesson Summary

The volume formula for a right square pyramid is $V = \frac{1}{3}Bh$, where B is the area of the square base.

The lateral length of a cone, sometimes referred to as the slant height, is the side s, shown in the diagram below.

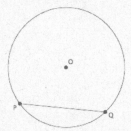

Given the lateral length and the length of the radius, the Pythagorean theorem can be used to determine the height of the cone.

Let O be the center of a circle, and let P and Q be two points on the circle. Then $\overline{PQ}$ is called a chord of the circle.

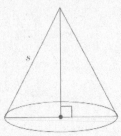

The segments OP and OQ are equal in length because both represent the radius of the circle. If the angle formed by POQ is a right angle, then the Pythagorean theorem can be used to determine the length of the radius when given the length of the chord, or the length of the chord can be determined if given the length of the radius.

EUREKA
MATH

Name _____   Date _____

Which has the larger volume? Give an approximate answer rounded to the tenths place.

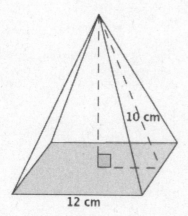

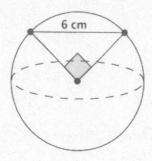

1. What is the lateral length of the cone shown below? Give an approximate answer rounded to the tenths place.

Let c in. be the lateral length in inches.

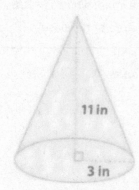

11 in

3 in

$$11^2 + 3^2 = c^2$$
$$121 + 9 = c^2$$
$$130 = c^2$$
$$\sqrt{130} = \sqrt{c^2}$$
$$\sqrt{130} = c$$

> I know the slanted part of the cone is called the lateral length. I can use the Pythagorean theorem to find its length.

The number $\sqrt{130}$ is between 11 and 12. In the sequence of tenths, it is between 11.4 and 11.5. Since 130 is closer to 11.4^2 than 11.5^2, the approximate value of the number is 11.4.

The lateral length of the cone is approximately 11.4 in.

2. The cone below has a base radius of 6 cm in length and a lateral length of 10 cm. What is the volume of the cone? Give an exact answer.

Let h cm represent the height of the cone.

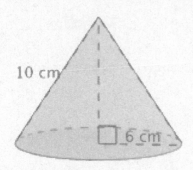

10 cm

6 cm

$$6^2 + h^2 = 10^2$$
$$36 + h^2 = 100$$
$$h^2 = 64$$
$$\sqrt{h^2} = \sqrt{64}$$
$$h = 8$$

The height of the cone is 8 cm.

$$V = \frac{1}{3}\pi(36)(8)$$
$$= 96\pi$$

> I can use the Pythagorean theorem to find the height of the cone. The volume of the cone is $\frac{1}{3}Bh$. B is the area of the base, which is a circle. I will not estimate π since the problem asked for an exact answer.

The volume of the cone is 96π cm^3.

EUREKA MATH

3. Determine the volume and surface area of the pyramid shown below. Give exact answers.

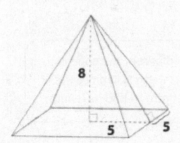

$$V = \frac{1}{3}(100)(8)$$
$$= \frac{800}{3}$$

The volume of a pyramid is the same as the volume of a cone, which is $\frac{1}{3}Bh$. B represents the area of the base.

The volume of the pyramid is $\frac{800}{3}$ cubic units.

Let c units represent the lateral length.

The base is the shape of a square with side lengths of 10. The four sides (faces) are the shape of triangles with base length of 10.

$$5^2 + 8^2 = c^2$$
$$25 + 64 = c^2$$
$$89 = c^2$$
$$\sqrt{89} = \sqrt{c^2}$$
$$\sqrt{89} = c$$

To find the surface area, I need to sum the areas of all the faces and the base of the pyramid.

The area of one face is $\frac{10\sqrt{89}}{2}$ square units, which is equal to $5\sqrt{89}$ square units. Since there are four faces, the total area is $4 \times 5\sqrt{89}$ square units, which is equal to $20\sqrt{89}$ square units.

The base area is 100 square units, and the total area of the faces is $20\sqrt{89}$ square units, so the surface area of the pyramid is $100 + 20\sqrt{89}$ square units.

4. What is the length of the chord of the sphere shown below? Give an exact answer using a square root.

Let c m represent the length of the chord.

$$6^2 + 6^2 = c^2$$
$$36 + 36 = c^2$$
$$72 = c^2$$
$$\sqrt{72} = \sqrt{c^2}$$
$$\sqrt{72} = c$$
$$\sqrt{6^2 \times 2} = c$$
$$6\sqrt{2} = c$$

A chord is a segment that connects any two points on a circle or sphere. The chord in the diagram is the hypotenuse of the right triangle.

6 m

The length of the chord is $6\sqrt{2}$ m.

The sides of the triangle are the same measurement since they are both radii of the sphere.

1. What is the lateral length (slant height) of the cone shown below? Give an approximate answer rounded to the tenths place.

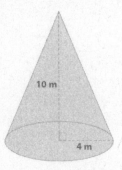

2. What is the volume of the cone shown below? Give an exact answer.

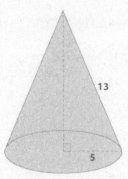

3. Determine the volume and surface area of the square pyramid shown below. Give exact answers.

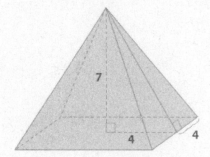

4. Alejandra computed the volume of the cone shown below as 64π cm^3. Her work is shown below. Is she correct? If not, explain what she did wrong, and calculate the correct volume of the cone. Give an exact answer.

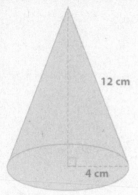

$$V = \frac{1}{3}\pi(4)^2(12)$$

$$= \frac{(16)(12)\pi}{3}$$

$$= 64\pi$$

The volume of the cone is 64π cm^3.

5. What is the length of the chord of the sphere shown below? Give an exact answer using a square root.

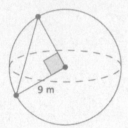

6. What is the volume of the sphere shown below? Give an exact answer using a square root.

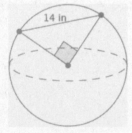

Lesson 19:   Cones and Spheres

EUREKA
MATH

Opening Exercise

Examine the bucket below. It has a height of 9 inches and a radius at the top of the bucket of 4 inches.

a. Describe the shape of the bucket. What is it similar to?

b. Estimate the volume of the bucket.

Example 1

Determine the volume of the truncated cone shown below.

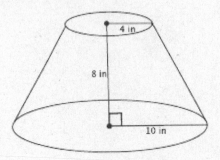

Exercises 1–5

1. Find the volume of the truncated cone.

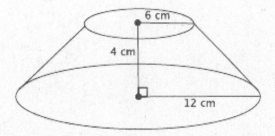

a. Write a proportion that will allow you to determine the height of the cone that has been removed. Explain what all parts of the proportion represent.

b. Solve your proportion to determine the height of the cone that has been removed.

c. Write an expression that can be used to determine the volume of the truncated cone. Explain what each part of the expression represents.

d. Calculate the volume of the truncated cone.

2. Find the volume of the truncated cone.

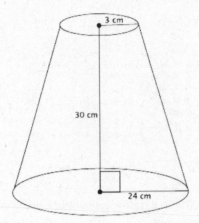

© 2019 Great Minds®. eureka-math.org

EUREKA
MATH®

3. Find the volume of the truncated pyramid with a square base.

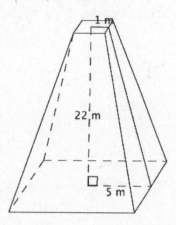

a. Write a proportion that will allow you to determine the height of the cone that has been removed. Explain what all parts of the proportion represent.

b. Solve your proportion to determine the height of the pyramid that has been removed.

c. Write an expression that can be used to determine the volume of the truncated pyramid. Explain what each part of the expression represents.

d. Calculate the volume of the truncated pyramid.

4. A pastry bag is a tool used to decorate cakes and cupcakes. Pastry bags take the form of a truncated cone when filled with icing. What is the volume of a pastry bag with a height of 6 inches, large radius of 2 inches, and small radius of 0.5 inches?

5. Explain in your own words what a truncated cone is and how to determine its volume.

EUREKA
MATH

Lesson Summary

A truncated cone or pyramid is the solid obtained by removing the top portion of a cone or a pyramid above a plane parallel to its base. Shown below on the left is a truncated cone. A truncated cone with the top portion still attached is shown below on the right.

Truncated cone: Truncated cone with top portion attached:

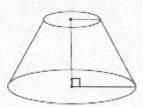

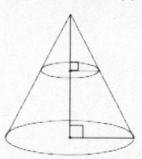

To determine the volume of a truncated cone, you must first determine the height of the portion of the cone that has been removed using ratios that represent the corresponding sides of the right triangles. Next, determine the volume of the portion of the cone that has been removed and the volume of the truncated cone with the top portion attached. Finally, subtract the volume of the cone that represents the portion that has been removed from the complete cone. The difference represents the volume of the truncated cone.

Pictorially,

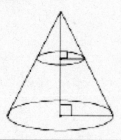 — =

EUREKA
MATH

© 2019 Great Minds®. eureka-math.org

Name _____   Date _____

Find the volume of the truncated cone.

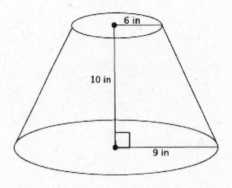

a. Write a proportion that will allow you to determine the height of
    the cone that has been removed. Explain what all parts of the
    proportion represent.

b. Solve your proportion to determine the height of the cone that
    has been removed.

c. Write an expression that can be used to determine the volume of the truncated cone. Explain what each part
    of the expression represents.

d. Calculate the volume of the truncated cone.

1. Find the volume of the truncated cone.

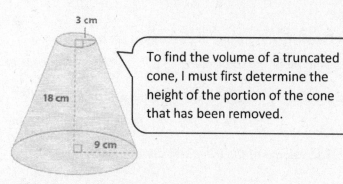

> To find the volume of a truncated cone, I must first determine the height of the portion of the cone that has been removed.

a. Write a proportion that will allow you to determine the height of the cone that has been removed. Explain what each part of the proportion represents.

$$\frac{3}{9} = \frac{x}{x + 18}$$

> I can use ratios that represent the corresponding sides of right triangles. Pictorially it looks like this:

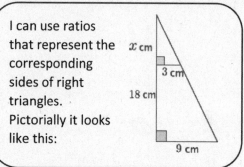

Let x cm represent the height of the small cone. Then x cm + 18 cm is the height of the larger cone. 3 cm is the base radius of the small cone, and 9 cm is the base radius of the large cone.

b. Solve your proportion to determine the height of the cone that has been removed.

$$3(x + 18) = 9x$$
$$3x + 54 = 9x$$
$$54 = 6x$$
$$9 = x$$

> This means the height of the cone before the smaller portion was removed was 27 cm, since $9 + 18 = 27$.

c. Show a fact about the volume of the truncated cone using an expression. Explain what each part of the expression represents.

$$\frac{1}{3}\pi(9)^2(27) - \frac{1}{3}\pi(3)^2(9)$$

> I need to determine the difference between the volume of the large cone and the volume of the small cone.

The expression $\frac{1}{3}\pi(9)^2(27)$ represents the volume of the large cone, and $\frac{1}{3}\pi(3)^2(9)$ is the volume of the small cone. The difference in volumes gives the volume of the truncated cone.

d. Calculate the volume of the truncated cone.

Volume of the small cone:	Volume of the large cone:	Volume of the truncated cone:
$V = \frac{1}{3}\pi(3)^2(9)$	$V = \frac{1}{3}\pi(9)^2(27)$	$729\pi - 27\pi = 702\pi$
$= 27\pi$	$= 729\pi$	

The volume of the truncated cone is 702π cm^3.

> I can use the same process I used to determine the volume of a truncated cone to determine the volume of a truncated pyramid.

2. Find the volume of the truncated pyramid with a square base.

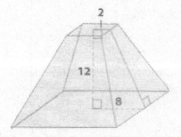

Let x units represent the height of the small pyramid.

$$\frac{2}{8} = \frac{x}{x+12}$$

$$2(x+12) = 8x$$

$$2x + 24 = 8x$$

$$24 = 6x$$

$$4 = x$$

> I learned how to determine the volumes of pyramids in the last lesson.

Volume of the small pyramid:	Volume of the large pyramid:	Volume of the truncated pyramid:
$V = \frac{1}{3}(16)(4)$	$V = \frac{1}{3}(256)(16)$	$\frac{4096}{3} - \frac{64}{3} = \frac{4032}{3}$
$= \frac{64}{3}$	$= \frac{4096}{3}$	

The volume of the truncated pyramid is $\frac{4032}{3}$ cubic units.

Lesson 20:   Truncated Cones

EUREKA
MATH

3. Challenge: Find the volume of the truncated cone.

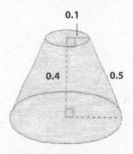

Since the height of the truncated cone is 0.4 units, we can drop a perpendicular line from the top of the cone to the bottom of the cone so that we have a right triangle with a leg length of 0.4 units and a hypotenuse of 0.5 units. By the Pythagorean theorem, if b units represents the length of the leg of the right triangle, then

> I need to determine the radius of the larger cone. I drew this picture to help:

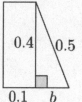

> To determine the radius of the larger cone, I will add $0.3 + 0.1 = 0.4$. The radius of the larger cone is 0.4 units.

$$0.4^2 + b^2 = 0.5^2$$
$$0.16 + b^2 = 0.25$$
$$b^2 = 0.09$$
$$b = 0.3.$$

The part of the radius of the bottom base found by the Pythagorean theorem is 0.3 units. When we add the length of the upper radius (because if you translate along the height of the truncated cone, it is equal to the remaining part of the lower base), the radius of the lower base is 0.4 units.

Let x units represent the height of the small cone.

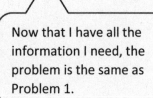

> Now that I have all the information I need, the problem is the same as Problem 1.

$$\frac{0.1}{0.4} = \frac{x}{x + 0.4}$$
$$0.1(x + 0.4) = 0.4x$$
$$0.1x + 0.04 = 0.4x$$
$$0.04 = 0.3x$$
$$\frac{0.04}{0.3} = x$$
$$\frac{4}{30} = \frac{2}{15} = x$$

> The height of the large cone is $\frac{2}{15} + 0.4$ which is equal to $\frac{8}{15}$.

Volume of the small cone:

$$V = \frac{1}{3}\pi(0.1)^2\left(\frac{2}{15}\right)$$
$$= \left(\frac{1}{3}\right)\left(\frac{1}{100}\right)\left(\frac{2}{15}\right)\pi$$
$$= \frac{2}{4500}\pi$$
$$= \frac{1}{2250}\pi$$

Volume of the large cone:

$$V = \frac{1}{3}\pi(0.4)^2\left(\frac{8}{15}\right)$$
$$= \left(\frac{1}{3}\right)\left(\frac{16}{100}\right)\left(\frac{8}{15}\right)\pi$$
$$= \frac{128}{4500}\pi$$
$$= \frac{64}{2250}\pi$$

Volume of the truncated cone:

$$V = \frac{64}{2250}\pi - \frac{1}{2250}\pi$$
$$= \left(\frac{64}{2250} - \frac{1}{2250}\right)\pi$$
$$= \frac{63}{2250}\pi$$

The volume of the truncated cone is $\frac{63}{2250}\pi$ cubic units.

1. Find the volume of the truncated cone.

    a. Write a proportion that will allow you to determine the height of the cone that has been removed. Explain what each part of the proportion represents.

    b. Solve your proportion to determine the height of the cone that has been removed.

    c. Show a fact about the volume of the truncated cone using an expression. Explain what each part of the expression represents.

    d. Calculate the volume of the truncated cone.

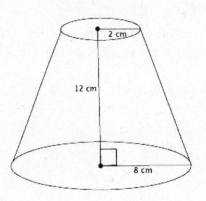

2. Find the volume of the truncated cone.

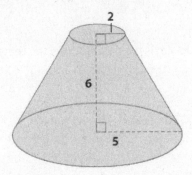

3. Find the volume of the truncated pyramid with a square base.

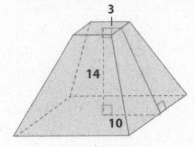

4. Find the volume of the truncated pyramid with a square base. Note: 3 mm is the distance from the center to the edge of the square at the top of the figure.

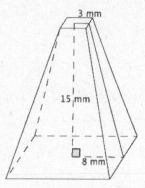

5. Find the volume of the truncated pyramid with a square base. Note: 0.5 cm is the distance from the center to the edge of the square at the top of the figure.

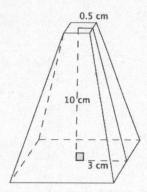

6. Explain how to find the volume of a truncated cone.

7. Challenge: Find the volume of the truncated cone.

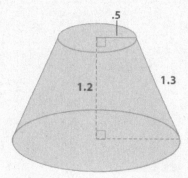

Lesson 20:   Truncated Cones

Exercises 1–4

1.

a. Write an expression that can be used to find the volume of the chest shown below. Explain what each part of your expression represents. (Assume the ends of the top portion of the chest are semicircular.)

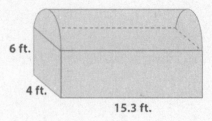

6 ft.

4 ft.

15.3 ft.

b. What is the approximate volume of the chest shown above? Use 3.14 for an approximation of π. Round your final answer to the tenths place.

2.

a. Write an expression for finding the volume of the figure, an ice cream cone and scoop, shown below. Explain what each part of your expression represents. (Assume the sphere just touches the base of the cone.)

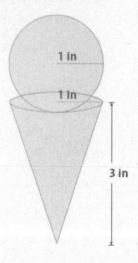

1 in

1 in

3 in

b. Assuming every part of the cone can be filled with ice cream, what is the exact and approximate volume of the cone and scoop? (Recall that exact answers are left in terms of π, and approximate answers use 3.14 for π). Round your approximate answer to the hundredths place.

3.

a. Write an expression for finding the volume of the figure shown below. Explain what each part of your expression represents.

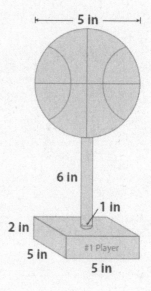

b. Every part of the trophy shown is solid and made out of silver. How much silver is used to produce one trophy? Give an exact and approximate answer rounded to the hundredths place.

4. Use the diagram of scoops below to answer parts (a) and (b).

   a. Order the scoops from least to greatest in terms of their volumes. Each scoop is measured in inches.
   (Assume the third scoop is hemi-spherical.)

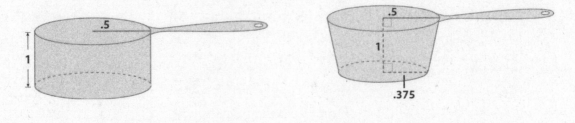

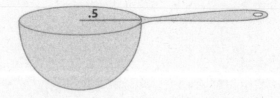

Lesson 21:   Volume of Composite Solids

© 2019 Great Minds®. eureka-math.org

EUREKA
MATH

b. How many of each scoop would be needed to add a half-cup of sugar to a cupcake mixture? (One-half cup is approximately 7 in^3.) Round your answer to a whole number of scoops.

Lesson Summary

Composite solids are figures comprising more than one solid. Volumes of composite solids can be added as long as no parts of the solids overlap. That is, they touch only at their boundaries.

Lesson 21:   Volume of Composite Solids

Name _____   Date _____

Andrew bought a new pencil like the one shown below on the left. He used the pencil every day in his math class for a week, and now his pencil looks like the one shown below on the right. How much of the pencil, in terms of volume, did he use?

Note: Figures are not drawn to scale.

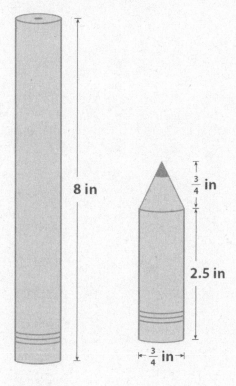

1. What volume of gel is required to completely fill up the lava lamp shown below? Note: 8 in. is the height of the truncated cone, not the lateral length of the cone.

    Let x in. represent the height of the portion of the cone that has been removed.

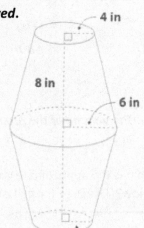

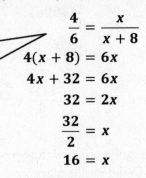

> I need to find the volume of the large cone and the volume of the cone that was removed like I did in the previous lesson.

$$\frac{4}{6} = \frac{x}{x+8}$$

$$4(x+8) = 6x$$

$$4x + 32 = 6x$$

$$32 = 2x$$

$$\frac{32}{2} = x$$

$$16 = x$$

Volume of the removed cone:

$$V = \frac{1}{3}\pi(4)^2(16)$$

$$= \frac{256}{3}\pi$$

Volume of the cone:

$$V = \frac{1}{3}\pi(6)^2(24)$$

$$= \frac{864}{3}\pi$$

Volume of one truncated cone:

$$\frac{864}{3}\pi - \frac{256}{3}\pi = \left(\frac{864}{3} - \frac{256}{3}\right)\pi$$

$$= \frac{608}{3}\pi$$

> There are two congruent truncated cones that make up the volume of the lava lamp. I need to multiply my result by 2 for the total volume.

The volume of gel needed to fill the lava lamp is $\frac{1216}{3}\pi$ in^3.

2.

    a. Write an expression for finding the volume of the prism with the pyramid portion removed. Explain what each part of your expression represents.

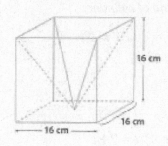

$$(16)^3 - \frac{1}{3}(16)^3$$

> I will subtract the volume of the pyramid from the volume of the cube.

The expression $(16)^3$ is the volume of the cube, and $\frac{1}{3}(16)^3$ is the volume of the pyramid. Since the pyramid's volume is being removed from the cube, we subtract the volume of the pyramid from the cube.

b. What is the volume of the prism shown above with the pyramid portion removed?

Volume of the prism:

$$V = (16)^3$$

$$= 4096$$

Volume of the pyramid:

$$V = \frac{1}{3}(4096)$$

$$= \frac{4096}{3}$$

I determined the volume by

$$4096 - \left(\frac{4096}{3}\right) = \frac{8192}{3}.$$

The volume of the prism with the pyramid removed is $\frac{8192}{3}$ cm³.

3. What is the approximate volume of the rectangular prism with two congruent cylindrical holes shown below? Use 3.14 for π. Round your answer to the tenths place.

The diagram shows the diameters of the cylindrical holes. The radius will be half the diameter. The height of the cylinder is the same as the width of the prism.

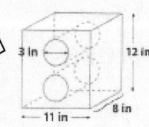

Volume of the prism:

$$V = (11)(8)(12)$$

$$= 1056$$

Volume of <u>one</u> cylinder:

$$V = \pi(1.5)^2(8)$$

$$= 18\pi$$

$$\approx 56.52$$

$$1056 - 56.52 - 56.52 = 942.96$$

The volume of the prism with the cylindrical holes is approximately 943 in³.

4. What is the exact total volume of the barbell shown below? Note: The two spheres are joined together by a cylinder.

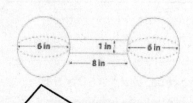

The diagram shows the diameters of the spheres. The radius will be half the diameter.

Volume of each sphere:

$$V = \frac{4}{3}\pi(3)^3$$

$$= 36\pi$$

Volume of cylinder:

$$V = (0.5)^2\pi(8)$$

$$= 2\pi$$

$$36\pi + 36\pi + 2\pi = (36 + 36 + 2)\pi = 74\pi$$

The total volume of the barbell is 74π in³.

EUREKA
MATH

1. What volume of sand is required to completely fill up the hourglass shown below? Note: 12 m is the height of the truncated cone, not the lateral length of the cone.

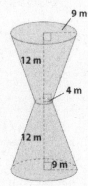

2.

   a. Write an expression for finding the volume of the prism with the pyramid portion removed. Explain what each part of your expression represents.

   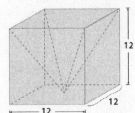

   b. What is the volume of the prism shown above with the pyramid portion removed?

3.

   a. Write an expression for finding the volume of the funnel shown to the right. Explain what each part of your expression represents.

   b. Determine the exact volume of the funnel.

   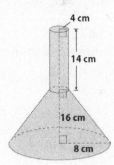

4. What is the approximate volume of the rectangular prism with a cylindrical hole shown below? Use 3.14 for π.
    Round your answer to the tenths place.

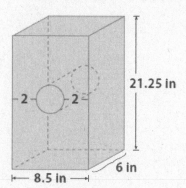

21.25 in

2   2

6 in

8.5 in

5. A layered cake is being made to celebrate the end of the school year. What is the exact total volume of the cake
    shown below?

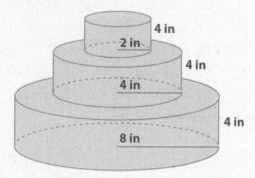

4 in

2 in

4 in

4 in

4 in

8 in

EUREKA
MATH

Exercise

The height of a container in the shape of a circular cone is 7.5 ft., and the radius of its base is 3 ft., as shown. What is the total volume of the cone?

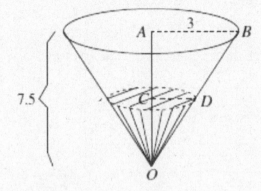

Time (in minutes)	Water Level (in feet)
	1
	2
	3
	4
	5
	6
	7
	7.5

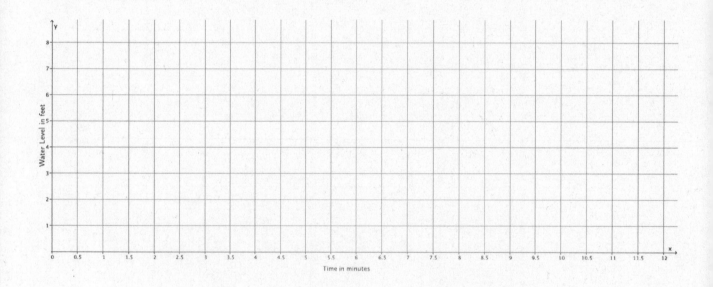

Lesson 22:   Average Rate of Change

EUREKA
MATH®

Name _____   Date _____

A container in the shape of a square base pyramid has a height of 5 ft. and a base length of 5 ft., as shown. Water flows into the container (in its inverted position) at a constant rate of 4 ft³ per minute. Calculate how many minutes it would take to fill the cone at 1 ft. intervals. Organize your data in the table below.

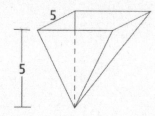

Water Level (in feet)	Area of Base (in feet²)	Volume (in feet³)	Time (in minutes)
1			
2			
3			
4			
5			

a.   How long will it take to fill up the container?

b.   Show that the water level is not rising at a constant rate. Explain.

1. Complete the table below for more intervals of water levels of the cone discussed in class. Then, graph the data on a coordinate plane.

Time (in minutes)	Water Level (in feet)
0.028	1
0.055	1.25
0.15	1.75
0.22	2
0.32	2.25
0.58	2.75
0.75	3
1.78	4
3.49	5

> I already know some of the times it takes to fill the cone from the work I did in class.

> I will use the proportion $\dfrac{3}{\text{radius}} = \dfrac{7.5}{\text{water level}}$ to find the radius like we did in class. Next, I will determine the volume of the cone with the radius I just determined. I will divide this amount by the rate, 6 ft^3 per minute, at which the cone is being filled to give me the time it takes to fill the cone to the given heights.

> I will add to the graph we started in class.

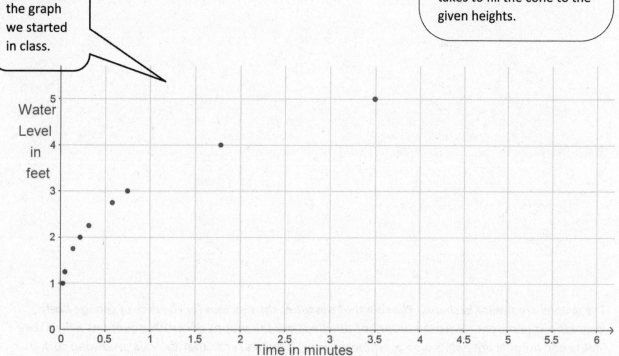

2. Complete the table below, and graph the data on a coordinate plane. Compare the graphs from Problems 1 and 2. What do you notice? If you could write a rule to describe the function of the rate of change of the water level of the cone, what might the rule include?

The inputs are all perfect squares. I need to add 3 to the result of $\sqrt{x}$ for each of the outputs.

x	$\sqrt{x} + 3$
1	4
4	5
9	6
16	7
25	8
36	9
49	10
64	11

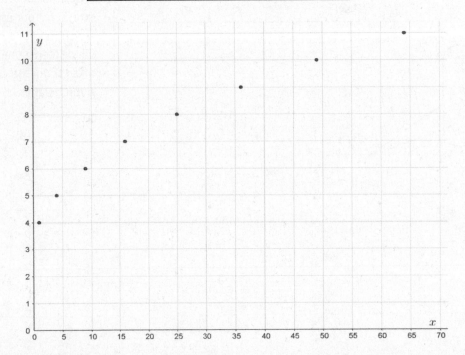

The graphs are similar in shape. The rule that describes the function for the rate of change likely includes a square root. Since the graphs of functions are the graphs of certain equations where their inputs and outputs are points on a coordinate plane, it makes sense that the rule producing such a curve would be a graph of some kind of square root.

Lesson 22:   Average Rate of Change

EUREKA
MATH

1. Complete the table below for more intervals of water levels of the cone discussed in class. Then, graph the data on a coordinate plane.

Time (in minutes)	Water Level (in feet)
	1
	1.5
	2
	2.5
	3
	3.5
	4
	4.5
	5
	5.5
	6
	6.5
	7
	7.5

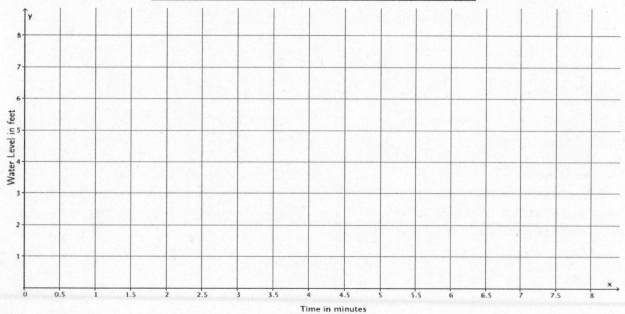

2. Complete the table below, and graph the data on a coordinate plane. Compare the graphs from Problems 1 and 2. What do you notice? If you could write a rule to describe the function of the rate of change of the water level of the cone, what might the rule include?

x	$\sqrt{x}$
1	
4	
9	
16	
25	
36	
49	
64	
81	
100	

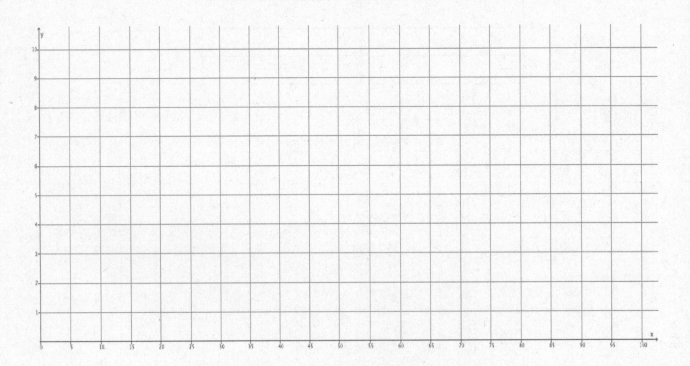

Lesson 22:   Average Rate of Change

EUREKA MATH

3. Describe, intuitively, the rate of change of the water level if the container being filled were a cylinder. Would we get the same results as with the cone? Why or why not? Sketch a graph of what filling the cylinder might look like, and explain how the graph relates to your answer.

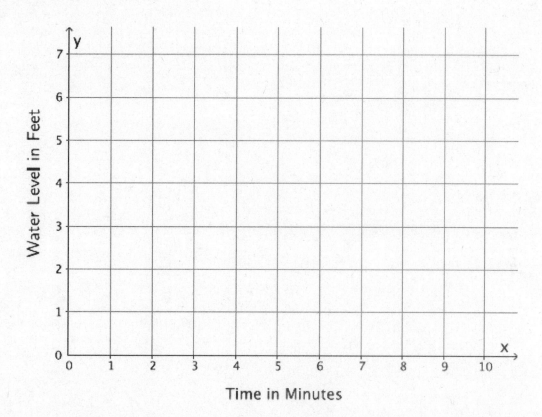

4. Describe, intuitively, the rate of change if the container being filled were a sphere. Would we get the same results as with the cone? Why or why not?

Mathematical Modeling Exercise

A ladder of length L ft. leaning against a wall is sliding down. The ladder starts off being flush with (right up against) the wall. The top of the ladder slides down the vertical wall at a constant speed of v ft. per second. Let the ladder in the position L_1 slide down to position L_2 after 1 second, as shown below.

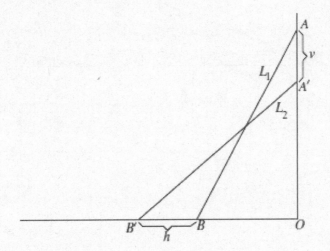

Will the bottom of the ladder move at a constant rate away from point O?

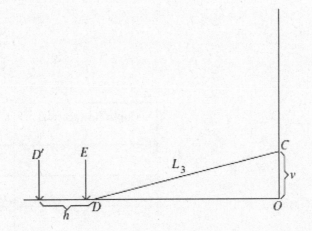

Consider the three right triangles shown below, specifically the change in the length of the base as the height decreases in increments of 1 ft.

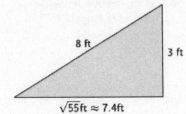

8 ft

3 ft

$\sqrt{55}$ft ≈ 7.4ft

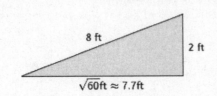

8 ft

2 ft

$\sqrt{60}$ft ≈ 7.7ft

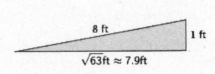

8 ft

1 ft

$\sqrt{63}$ft ≈ 7.9ft

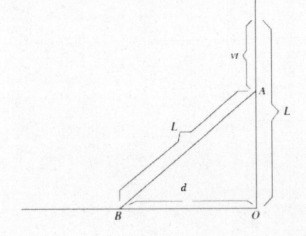

Input (in seconds) t	Output (in feet) $d = \sqrt{225 - (15 - t)^2}$
0	
1	
3	
4	
7	
8	
14	
15	

Lesson 23:   Nonlinear Motion

EUREKA
MATH

Name _____   Date _____

Suppose a book is 5.5 inches long and leaning on a shelf. The top of the book is sliding down the shelf at a rate of 0.5 in. per second. Complete the table below. Then, compute the average rate of change in the position of the bottom of the book over the intervals of time from 0 to 1 second and 10 to 11 seconds. How do you interpret these numbers?

Input (in seconds) t	Output (in inches) $d = \sqrt{30.25 - (5.5 - 0.5t)^2}$
0	
1	
2	
3	
4	
5	
6	
7	
8	
9	
10	
11	

1. Suppose a ladder is 12 feet long, and the top of the ladder is sliding down the wall at a rate of 0.9 ft. per second. Compute the average rate of change in the position of the bottom of the ladder over the intervals of time from 0 to 0.5 seconds, 3 to 3.5 seconds, 7 to 7.5 seconds, 9.5 to 10 seconds, and 12 to 12.5 seconds. How do you interpret these numbers?

> I can use a calculator to help with the substitution of the input, t, into the equation to get the output, d.

Input t	Output $d = \sqrt{144 - (12 - 0.9t)^2}$
0	$\sqrt{0} = 0$
0.5	$\sqrt{10.60} \approx 3.26$
3	$\sqrt{57.51} \approx 7.58$
3.5	$\sqrt{65.68} \approx 8.1$
7	$\sqrt{111.51} \approx 10.56$
7.5	$\sqrt{116.44} \approx 10.79$
9.5	$\sqrt{132.1} \approx 11.49$
10	$\sqrt{135} \approx 11.62$
12	$\sqrt{142.56} \approx 11.94$
12.5	$\sqrt{143.44} \approx 11.98$

> We did an example like this in class where d represented the distance from the bottom of the ladder to the corner where the wall intersected the floor.

The average rate of change between 0 and 0.5 seconds is

$$\frac{3.26 - 0}{0.5 - 0} = \frac{3.26}{0.5} = 6.52.$$

The average rate of change between 3 and 3.5 seconds is

$$\frac{8.1 - 7.58}{3.5 - 3} = \frac{0.52}{0.5} = 1.04.$$

The average rate of change between 7 and 7.5 seconds is

$$\frac{10.79 - 10.56}{7.5 - 7} = \frac{0.23}{0.5} = 0.46.$$

> The average rates are not the same over the time intervals. This means that the ladder is not sliding down the wall at a constant rate. This means the average rate of change is not linear.

The average rate of change between 9.5 and 10 seconds is

$$\frac{11.62 - 11.49}{10 - 9.5} = \frac{0.13}{0.5} = 0.26.$$

The average rate of change between 12 and 12.5 seconds is

$$\frac{11.98 - 11.94}{12.5 - 12} = \frac{0.04}{0.5} = 0.08.$$

> The average rates are getting smaller as the ladder slides down the wall.

The average speeds show that the rate of change in the position of the bottom of the ladder is not linear. Furthermore, it shows that the rate of change in the position at the bottom of the ladder is quick at first, 6.52 feet per second in the first half second of motion, and then slows down to 0.08 feet per second in the half second interval from 12 to 12.5 seconds.

2. Will any length of ladder, L, and any constant speed of sliding of the top of the ladder v ft. per second, ever produce a constant rate of change in the position of the bottom of the ladder? Explain.

No, the rate of change in the position at the bottom of the ladder will never be constant. We showed that if the rate were constant, there would be movement in the last second of the ladder sliding down that wall that would place the ladder in an impossible location. That is, if the rate of change were constant, then the bottom of the ladder would be in a location that exceeds the length of the ladder. Also, we determined that the distance that the bottom of the ladder is from the wall over any time period can be found using the formula $d = \sqrt{L^2 - (L - vt)^2}$, which is a nonlinear equation. Since graphs of functions are equal to the graph of a certain equation, the graph of the function represented by the formula $d = \sqrt{L^2 - (L - vt)^2}$ is not a line, and the rate of change in position at the bottom of the ladder is not constant.

> I tried this experiment with a book, and I could see that the book slid down the wall quickly at first but slowed down when it hit the floor.

> In class, we learned that the equation $d = \sqrt{L^2 - (L - vt)^2}$ is not linear. This means the average rates of change will not be constant.

EUREKA MATH

1. Suppose the ladder is 10 feet long, and the top of the ladder is sliding down the wall at a rate of 0.8 ft. per second. Compute the average rate of change in the position of the bottom of the ladder over the intervals of time from 0 to 0.5 seconds, 3 to 3.5 seconds, 7 to 7.5 seconds, 9.5 to 10 seconds, and 12 to 12.5 seconds. How do you interpret these numbers?

Input (in seconds) t	Output (in feet) $d = \sqrt{100 - (10 - 0.8t)^2}$
0	
0.5	
3	
3.5	
7	
7.5	
9.5	
10	
12	
12.5	

2. Will any length of ladder, L, and any constant speed of sliding of the top of the ladder, v ft. per second, ever produce a constant rate of change in the position of the bottom of the ladder? Explain.

Credits

Great Minds® has made every effort to obtain permission for the reprinting of all copyrighted material. If any owner of copyrighted material is not acknowledged herein, please contact Great Minds for proper acknowledgment in all future editions and reprints of this module.